Lecture Notes in Control and Information Sciences

Edited by M. Thoma and A. Wyner

Vol. 81: Stochastic Optimization
Proceedings of the International Conference, Kiew, 1984
Edited by I. Arkin, A. Shiraev, R. Wets
X, 754 pages, 1986.

Vol. 82: Analysis and Algorithms of Optimization Problems
Edited by K. Malanowski, K. Mizukami
VIII, 240 pages, 1986.

Vol. 83: Analysis and Optimization of Systems
Proceedings of the Seventh International Conference of Analysis and Optimization of Systems
Antiba, June 26-27, 1986
Edited by A. Bensoussan, J. L. Lions
XVI, 901 pages, 1986.

Vol. 84: System Modelling and Optimization
Proceedings of the 12th IFIP Conference
Budapest, Hungary, September 2–6, 1985
Edited by A. Prékopa, J. Szelezsán, B. Strazicky
XII, 1046 pages, 1986.

Vol. 85: Stochastic Processes in Underwater Acoustics
Edited by Charles R. Baker
V, 205 pages, 1986.

Vol. 86: Time Series and Linear Systems
Edited by Sergio Bittanti
XVII, 243 pages, 1986.

Vol. 87: Recent Advances in System Modelling and Optimization
Proceedings of the IFIP-WG 7/1 Working Conference
Santiago, Chile, August 27-31, 1984
Edited by L. Contesse, R. Correa, A. Weintraub
IV, 199 pages, 1987.

Vol. 88: Bruce A. Francis
A Course in H_∞ Control Theory
XI, 156 pages, 1987.

Vol. 88: Bruce A. Francis
A Course in H_∞ Control Theory
X, 150 pages, 1987.
Corrected - 1st printing 1987

Vol. 89: G. K. H. Pang/A. G. J. McFarlane
An Expert System Approach to Computer-Aided Design of Multivariable Systems
XII, 223 pages, 1987.

Vol. 90: Singular Perturbations and Asymptotic Analysis in Control Systems
Edited by P. Kokotovic, A. Bensoussan, G. Blankenship
VI, 419 pages, 1987.

Vol. 91 Stochastic Modelling and Filtering
Proceedings of the IFIP-WG 7/1 Working Conference
Rome, Italy, Decembre 10-14, 1984
Edited by A. Germani
IV, 209 pages, 1987.

Vol. 92: L. T. Grujić, A. A. Martynyuk, M. Ribbens-Pavella
Large-Scale Systems Stability Under Structural and Singular Perturbations
XV, 366 pages, 1987.

Vol. 93: K. Malanowski
Stability of Solutions to Convex Problems of Optimization
IX, 137 pages, 1987.

Vol. 94: H. Krishna
Computational Complexity of Bilinear Forms
Algebraic Coding Theory and Applications to Digital Communication Systems
XVIII, 166 pages, 1987.

Vol. 95: Optimal Control
Proceedings of the Conference on Optimal Control and Variational Calculus
Oberwolfach, West-Germany, June 15-21, 1986
Edited by R. Bulirsch, A. Miele, J. Stoer and K. H. Well
XII, 321 pages, 1987.

Vol. 96: H. J. Engelbert/W. Schmidt
Stochastic Differential Systems
Proceedings of the IFIP-WG 7/1 Working Conference
Eisenach, GDR, April 6-13, 1986
XII, 381 pages, 1987.

For information about Vols. 1–80 please contact your bookseller or Springer-Verlag.

Lecture Notes in
Control and
Information Sciences
Edited by M. Thoma and A. Wyner

155

M. Bouvet, G. Bienvenu (Eds.)

High-Resolution Methods in Underwater Acoustics

Springer-Verlag
Berlin Heidelberg GmbH

Editors
Michel Bouvet
Groupe d'Etudes et de Recherches
en Détection Sous-Marine
DCN Toulon
Le Brusc
83140 Six-Fours-Les-Plages
France

Georges Bienvenu
Thomson-Sintra Activités Sous-Marines
B. P. 138
Parc de Sophia-Antipolis
06561 Valbonne Cedex
France

ISBN 978-3-540-53716-8 ISBN 978-3-540-47013-7 (eBook)
DOI 10.1007/978-3-540-47013-7

Originally published by Springer-Verlag Berlin Heidelberg New York in 1991

61/3020-543210 Printed on acid-free paper.

HIGH-RESOLUTION METHODS IN UNDERWATER ACOUSTICS

Edited by

Michel BOUVET
Groupe d'Etudes et de Recherches en Détection Sous-Marine
DCN Toulon (France)

Georges BIENVENU
Thomson-Sintra ASM
Valbonne (France)

CONTRIBUTORS

Bhuyan A.
Tel Aviv University
TEL AVIV (Israel)

Bienvenu G.
Thomson-Sintra Activités Sous-Marines
B.P. 38, 06561 VALBONNE CEDEX (France)

Boashash B.
CRISSP, University of Queensland
BRISBANE 4067 (Australia)

Bouvet M.
Groupe d'Etudes et de Recherches en Détection Sous-Marine
DCN Toulon, Le Brusc, 83140 SIX-FOURS-LES-PLAGES (France)

Kirsteins I.P.
University of Rhode Island
KINGSTON, RI 02881 (USA)
(on training from Naval Underwater Systems Center
NEW LONDON, CT06320 (USA))

Le Cadre J.P.
Groupe d'Etudes et de Recherches en Détection Sous-Marine
DCN Toulon, Le Brusc, 83140 SIX-FOURS-LES-PLAGES (France)
(now with INSTITUT DE RECHERCHES EN INFORMATIQUE ET SYSTEMES ALEATOIRES
Campus Universitaire de Beaulieu, 35042 RENNES CEDEX (France))

Messer H.
Tel Aviv University
TEL AVIV (Israel)

Nicolas P.
SACLANT Undersea Research Center
LA SPEZIA, 19026 (Italie)

Owsley N.L.
Naval Underwater Systems Center
NEW LONDON, CT 06320 (USA)

Ravazzola P.
Groupe d'Etudes et de Recherches en Détection Sous-Marine
DCN Toulon, Le Brusc, 83140 SIX-FOURS-LES-PLAGES (France)
(now with AERO, 3, Avenue de l'Opéra, 75001 PARIS (France))

Schultheiss P.M.
Yale University
Yale Station, NEW HAVEN, CT 06520-2157 (USA)

Speiser J.M.
Naval Ocean Systems Center
SAN DIEGO, CA 92152-5000 (USA)

Tufts D.W.
University of Rhode Island
KINGSTON, RI 02881 (USA)

Vezzosi G.
Thomson-Sintra Activités Sous-Marines
B.P. 38, 06561 VALBONNE CEDEX (France)
(now with INSTITUT DE RECHERCHES EN INFORMATIQUE ET SYSTEMES ALEATOIRES
Campus Universitaire de Beaulieu, 35042 RENNES CEDEX (France))

Whitehouse H.J.
Naval Ocean Systems Center
SAN DIEGO, CA 92152-5000 (USA)

INTRODUCTION

Michel BOUVET
Groupe d'Etudes et de Recherches en Détection Sous-Marine
(France)

Georges BIENVENU
Thomson-Sintra ASM
(France)

INTRODUCTION

UNDERWATER ACOUSTICS AND HIGH-RESOLUTION

One of the most challenging problem in electrical engineering is the detection and estimation of targets in underwater acoustics. Solutions to this problem use tools given by signal processing techniques and, in particular, by the so-called high-resolution methods. Their fundamental philosophy is to overcome the limitations of the classical Fourier transform in terms of resolution. By resolution, we mean the separation of two close sources with same power. The two main subjects using Fourier transform are spectral analysis and array processing.

The high-resolution idea has lead to a tremendous, and somehow brownian, activity expressed by a huge number of publications, both in reviews and conferences, since the pioneering works of Pisarenko, Ligget and Mermoz in the mid-seventies. It is worth noting that mathematical background was given by the works of Caratheodory in 1911. Applications in underwater acoustics have appeared at the end of the seventies.

The main idea is to assume a parametric model generating the observation. The limitation of the classical methods, based on the Fourier transform, is due to the necessary truncation of the observation window. In spectral analysis, the observation is available for a limited time. In array processing, the array has a limited extension. By using a parametric model, it is possible to put values on the observation out of this window (or pupils) and, then, to "artificially" extend them. This has lead to the so-called Maximum Entropy methods.

Another approach consists in modelling more precisely the noise structure and in using more a priori information. This is the philosophy of most of the high-resolution methods that will be studied in the sequel.

SIGNAL PROCESSING

With respect to other fields, underwater acoustics signal processing has some specific characteristics.

* Compared to radar, it deals with a more severe noise environment, in terms of non-stationarities, power level, power fluctuations, propagation phenomenon, ... In counterpart, due to the rather slow sound celerity, more time is available. More complex algorithms can then be used.

* Compared to communications, the main difference comes from the lack of knowledge on the signal which has to be detected, especially in passive listening. The detection problem is harder.

* Compared to speech processing, signals are more stationary. This is the reason why, in underwater acoustics, high-resolution methods seem to be more useful in array processing than in spectral analysis.

In the time domain, speech processing algorithms can only use short data records because of small stationary time. High-resolution methods will then be used in the time-frequency transformation (formants analysis). In underwater acoustics, the quasi-stationarity time is great enough to neglect the drawbacks of Fourier transform. On the contrary, it is useful to separate spatially two objects using an array which is, for geometrical and practical reasons, limited. **The main application field for high-resolution methods in sonar is array processing.**

Let us note here that such methods can be used for other purposes like separation of multipath in elevation, estimation of highlights, ... But the emphasis will be put on array processing: estimation of the direction of sources in a 2-D space.

For various reasons, no hardware realization of high-resolution methods has yet been made in underwater acoustic. It seems that it is time to perform a real implementation of such methods for sonar array processing. This evolution is allowed by a tremendous improvement in electronic capacity in terms of speed of computation and memory access, of memory size and of ability to implement the complex algorithms that are needed, like eigen or singular value decomposition.

PHILOSOPHY AND NOTATION

Let us assume that we have an array of N sensors. The output of the i-th sensor being denoted by $x_i(t)$, t being the time, the covariance matrix of the received observation is $R(t) = E(\underline{x}(t)\,\underline{x}^*(t))$, $\underline{x}(t)$ is the vector containing the $x_i(t)$ as elements, $\underline{x}(t) = (x_1(t), x_2(t), ..., x_N(t))^t$. Transposition is denoted by the superscipt $.^t$, transposition-conjugation by $.^*$ and statistical expectation by E. The interspectral matrix $\Gamma(f)$ is the Fourier transform of R(t). For a single source without noise, we have $\Gamma(f) = \gamma(f)\,\underline{d}(f)\,\underline{d}^*(f)$, where $\gamma(f)$ is the spectral density of the received signal from this source and where $\underline{d}(f)$ is the direction vector. This last vector is composed of the N transfert functions between the source and each of the sensors, normalized by the transfert function between the source and a reference on the array.

In the case of M sources in noise, the interspectral matrix can be expressed as $\Gamma(f) = \sum_{i=1}^{m} \gamma_i(f)\,\underline{d}_i(f)\,\underline{d}_i^*(f) + \Gamma_n(f)$ where $\Gamma_n(f)$ is the noise interspectral matrix. The power of the high-resolution methods comes form the separation of $\Gamma(f)$ into two parts:

* a **signal** part, $\Gamma_s(f) = \sum_{i=1}^{m} \gamma_i(f)\,\underline{d}_i(f)\,\underline{d}_i^*(f)$

* a **noise** part, $\Gamma_n(f)$.

Asymptotically (as observation time goes to infinity), this separation is not limited by the signal-to-noise ratio or the closeness of the two sources.

ORGANIZATION

The first chapter, "***High-resolution passive array processing: an overview of principles***", by G. Bienvenu and N.L. Owsley, presents the basic ideas upon which the high-resolution methods are based. It is an overview of their principles and their properties. It discusses their performances bounds and other important topics such as the estimation of the number of sources and the extension to wideband signals.

If the resolution of such methods is not theoretically limited, it is practically limited by the lack of knowledge on the exact model, of $\Gamma_s(f)$ and $\Gamma_n(f)$ in particular. The second chapter, "***Practical use of parametric methods for array processing***", by J.P. Le Cadre and P. Ravazzola, gives a way to overcome the usual white noise hypothesis, or, more generally, a noise model misadjustment.The techniques presented are twofolds: first, estimate the noise model, adaptively; and, second, use more fundamentally the plane wave hypothesis.

P. Nicolas and G. Vezzosi in chapter 3, "***Localization of broad-band sources with an array of unknown geometry***", solve a very difficult problem: the use of an array of unknown geometry in order to estimate the direction of arrival of broadband sources. They show that, under usual hypothesis, three sources are enough to estimate, in the two dimensionnal case, the sensor locations and the source bearings, at the same time.

In chapter 4, "***High-resolution processing techniques for temporal and spatial signals***", H.J. Whitehouse, B. Boashash and J.M. Speiser put emphasis on high-resolution methods for temporal analysis of signals and, especially, non-stationary signals. Going in another "space", they present the Wigner-Ville distribution whose idea is different from the bearing estimation methods presented in the other chapters of this volume, but whose results are similar: to obtain a better "resolution" and caracterization than with classical Fourier methods.

Chapter 5, "***Performance limits of high-resolution systems***", by A. Bhuyan, H. Messer and P.M. Schultheiss, addresses the problem of sensitivity of high-resolution direction finding methods to uncertainties on several factors such as array geometry or propagation conditions. In realistic situations, knowledge of sensor positions or parameters like signal or noise levels is unperfect. Nevertheless, it is possible to compute performance bounds. This chapter is devoted to the computation of Cramer-Rao lower bounds in several situations.

In some real situations, the weak performances of the classical Fourier based methods in terms of resolution are due to the presence of interference. The work, done by I.P. Kirsteins and D.W. Tufts and presented in chapter 6, "***Rapidly adaptive nulling of interference***", describes a method able to deal with interference, namely to suppress it using a rapidly adaptive algorithm. It contains theoretical performance computations as well as simulation results and real data analysis.

AIMS

The principal aim of this book is to make a synthesis of results related to the use of high-resolution methods in underwater acoustics. It is intented to give the state of the art to the interested people. This book can be used as :

*a text-book for the university community (professors, PhD students);
*a reference book for researchers and industry engineers, possibly from fields other than underwater acoustics;
*decisional elements for "operational people".

This has been obtained by the mean of research papers written by experts on different subjects going from theory (synthesis of methods, theoretical bounds of performances, ...) to application (implementation, at-sea results, ...).

ACKNOWLEGMENTS

Earlier version of the works contained in this volume have been presented during a workshop associated with the GRETSI Conference, in Juan-les-Pins (France), in 12-16 June 1989.

Let us thank the GRETSI (Groupe d'Etudes et de Recherches en Traitement du Signal et des Images) and the DRET (Direction des Recherches et Etudes Techniques de la Délégation Générale pour l'Armement) for their financial support and the help they have provided for the organization of the workshop. Above all, we are very gratefull to all the authors for their contribution to this volume. Without them, this book would not have existed.

CHAPTER 1

HIGH-RESOLUTION PASSIVE ARRAY PROCESSING: AN OVERVIEW OF PRINCIPLES

G. BIENVENU
Thomson-Sintra ASM, Valbonne (France)

N.L. OWSLEY
Naval Underwater Systems Center, New London (USA)

HIGH RESOLUTION PASSIVE ARRAY PROCESSING : AN OVERVIEW OF PRINCIPLES

Georges BIENVENU* and Norman L. OWSLEY**

* THOMSON SINTRA ASM BP 38 06561 VALBONNE CEDEX -FRANCE-
** NAVAL UNDERWATER SYSTEMS CENTER NEW-LONDON, CT. 06320 -USA-

SUMMARY

High resolution array processing has been an important field of interest. This is due to the fact that often, and this is particularly true for passive sonar, the array aperture is small, which results in a weak resolving power, and also secondary lobes are difficult to maintain at a sufficiently low-level to reduce the masking effect due to high level jammers.
This paper presents an overview of the principles of high resolution passive array processing and of their properties. Performance bouds of array processing are examined, which show that resolving power can be enhanced by increasing the observation time. Resolving power of high resolution methods has the same behaviour, but with a lower gain. High resolution methods beamforming needs an estimate of the number of sources : a unified method is presented which simplifies this problem. Extension to wideband signals is presented and results obtained on real date are shown.

1. INTRODUCTION

Array processing performance is a key point in passive sonar. It is the function which allows the separation in the bearing space of the different noise sources which are in the medium. In some sense, the ideal processing should build the sharpest image as possible of space arround the receiving array, just as a telescope or an eye does. The basic processing used in sonar is the conventional beamformer, which is an electronic copy of the function performed by an optical lense.
As any processing, array processing is based on a model of the physical problem under consideration. In passive sonar, the problem to handle is detection of ships surrounded by background noise and estimation of their specific paramaters. These sources are always far enough from the array compared to its size to be considered as ponctual. Conventional beamforming uses only a model of the propagation between the sources of interest and the array. It assumes that the sources propagate coherent wavefronts on the sensors. They are said to have a perfect spatial coherence : the signals received from a source on the sensors are identical except for a linear filtering. It assumes also that the spatial shape of the wavefront received on the array from a source is a known function of the source position $\underline{\theta}$. With such a model, the cross-spectral density matrix (CSDM) of the signals received on the sensors for a single source is equal to :

$$\Gamma(f) = \gamma(f)\ \underline{d}(f,\underline{\theta})\ \underline{d}^{+}(f,\underline{\theta}) \tag{1}$$

(+ denotes the conjugate transpose and $\underline{d}$ a column vector) where $\gamma(f)$ is the spectral density of the signal received on a reference point of the array, and $\underline{d}(f,\underline{\theta})$ the source position (or direction) vector composed of the transfer functions between each sensor and the reference point, including the transfer function of the sensors. The basic model is the pure time delay model :

$$\underline{d}(f,\underline{\theta}) = \{\exp[2\pi j f \tau_1(\underline{\theta}),\ldots, \exp[2\pi j f \tau_k(\underline{\theta}),\ldots, \exp[2\pi j f \tau_K(\underline{\theta})]\} \tag{2}$$

where $\tau_k(\underline{\theta})$ is the delay between sensor k and the reference point.
The principle of a beamforming function is given on fig. 1.
For each desired look direction $\underline{\theta}$, each sensor output is processed by a filter and then sumed with the others. The filters form the beamforming vector $\underline{H}(f,\underline{\theta})$.
The output spectral density is given by :

$$\gamma_c(f,\underline{\theta}) = \underline{H}^{+}(f,\underline{\theta})\ \Gamma(f)\ \underline{H}(f,\underline{\theta}) \tag{3}$$

If only one source is present in the direction $\underline{\theta}_s$, it comes :

$$\gamma_c(f,\underline{\theta}) = \gamma(f)\ |\underline{H}^{+}(f,\underline{\theta})\ \underline{d}(f,\underline{\theta}_s)|^2 \tag{4}$$

The function : $D(f,\underline{\theta},\underline{\theta}_s) = |\underline{H}^{+}(f,\underline{\theta})\ \underline{d}(f,\underline{\theta}_s)|^2$ is the directivity pattern (or ambiguity fuction) of the beamformer.
The performance of the conventional beamformer is given by this function, together with the gain against the background noise which is generally defined for a spherical isotropic distribution. The two main limitations are given by the 3 dB beamwidth of the mainlobe $2\theta_3$ which limits the resolving power (defined for two sources of equivalent levels) and the side-lobe level which limits the detection of a weak source in presence of high level sources (masking effect). The improvement of these features is the basic problem of array processing. For conventional beamforming, all the features are linked together. It is possible for example to decrease the side-lobe level in order to increase the protection against the jammers by using an appropriate shading. But the mainlobe will be broadened, which decreases the resolving power, and moreover the gain against the back-ground noise will be decreased.

2. LIMITS TO HIGH RESOLUTION ARRAY PROCESSING

The main axis which has been pursued in order to improve array processing performance is the improvement of resolving power. That is an estimation problem, for which a limit exists : performance cannot be increased indefinitely by processing.
Basically, it is limited by the array size, the structure of the noise field (the signal to noise ratio of the sources particularly), and more physically by the exact knowledge of the propagation in the medium, of the

sensor responses. A basic bound for the performance concerning array size and signal to noise ratio can be obtained in the case of an exact model using the Cramer-Rao Lower Bound. It states that for any unbiased estimator the covariance matrix V(f) of the parameter estimates is bounded by the inverse of the Fisher information matrix F(f) :

$V(f) - F^{-1}(f) \geq 0$. The p, q component of F(f) is equal to :

$$F_{pq}(f) = - \frac{\delta^2 L(f)}{\delta\beta_p \ \delta\beta_q}$$

where L(f) is the Log-likelihood function of the observation and β_p and β_q the p and q parameter components.
In the array processing case, the observation is a vector $\underline{X}(f)$ which is composed of the Fourier transforms over a time length T of the signals received on the sensors. These signals are assumed to be jointly Gaussian with zero mean. If T is large enough, its covariance matrix can be considered equal to the spectral density matrix $\Gamma(f)$. If the noise field is composed of P independant sources, and of incoherent backgroundnoise, $\Gamma(f)$ is equal to :

$$\Gamma(f) = \sum_{i=1}^{P} \gamma_i(f) \ \underline{d}(f,\underline{\theta}_i) \ \underline{d}^+(f,\underline{\theta}_i) + \sigma(f) \ I \tag{5}$$

where $\gamma_i(f)$ is the spectral density of source i and $\underline{d}(f,\underline{\theta}_i)$ its position vector, $\sigma(f)$ is the spectral density of the background noise and I the identity matrix. If the observation consists of N independent observation vectors $\underline{X}_n(f)$, the Log-likelihood function is equal to :

$$L(f) = -KN \operatorname{Log} \pi - N \operatorname{Log} |\Gamma(f)| - N\operatorname{tr}[\widehat{\Gamma}(f) \ \Gamma^{-1}(f)] \tag{6}$$

where $|\Gamma(f)|$ is the determinant of $\Gamma(f)$, tr stands for trace and $\widehat{\Gamma}(f)$ is equal to :

$$\frac{1}{N} \sum_{n=1}^{N} \underline{X}_n(f)^+ \ \underline{X}_n(f) \tag{7}$$

It can be shown that in the general case the Fisher matrix components are given by :

$$F_{pq}(f) = N\operatorname{tr} \left[\Gamma^{-1}(f) \frac{\partial\Gamma(f)}{\partial\beta_p} \Gamma^{-1}(f) \frac{\partial\Gamma(f)}{\partial\beta_q}\right]$$

The unknown parameters to be estimated are $\gamma_i(f)$, $\sigma(f)$ and $\underline{\theta}_i$. The components of the Fisher matrix can be computed [1, 2]. The covariance matrix of the parameter estimates has the remarkable property to be inversely proportional to the number of observations N, thus as N goes to

infinity, any parameter can be estimated perfectly.
Particularly, concerning the resolving power, though it is not directly given by the Cramer-Rao bound, its behaviour can be appraciated by the quantity $\sigma_{\hat{\varepsilon}/\varepsilon}$, where $\sigma_{\hat{\varepsilon}}$ is the standard deviation of the estimate $\hat{\varepsilon}$ of the separation ε between two equal level sources : this quantity is inversely proportional to $\sqrt{N}$. Therefore, the possibility exists of finding a processing which can resolve two sources as weak and close together they may be by increasing the observation time.
Morever, the resolving power also increases with the signal to noise ratio. That can be seen on an example with a linear array of 9 equispaced sensors: fig. 2 shows the signal to noise ratio (10 log $[\gamma(f)/\sigma(f)]$ versus $\varepsilon/2\theta_3$ for several values of $\sigma_{\hat{\varepsilon}/\varepsilon}$ and N = 100.

3. DATA DEPENDENT ARRAY PROCESSING METHODS

Different methods has been established in order to improve the performance of conventional beamforming. They have a common characteristic which is that the processing is continuously adapted to the noisefield using the received signals. The methods can be devided into two types : those whose resolving power is asymptotically limited, as adaptive beamforming (minimum variance) and maximum entropy method ; and those whose resolving power improves with the observation time without limit : the so-called high resolution methods.

3.1 Adaptive beamforming

As an example of the first type, adaptive beamforming performance will be described. It uses the same noisefield model as conventional beamforming. But unlike conventional beamforming, adaptive beamforming tries to minimize optimally the noise power whatever the noise field composition except a signal in the look-direction. It is derived from the maximum likelihood estimator of an unknown signal in a given look direction $\underline{d}(f,\underline{\theta})$. The corresponding spatial filter is given by (in a Gaussian context) :

$$\underline{H}(f,\underline{\theta}) = \Gamma_N^{-1}(f)\ \underline{d}(f,\underline{\theta})\ [\underline{d}^{+}(f,\underline{\theta})\ \Gamma_N^{-1}(f)\ \underline{d}(f,\underline{\theta})]^{-1} \qquad (8)$$

where $\Gamma_N(f)$ is the cross spectral density matrix of the noisefield except the source in the look-direction. It can be shown that the same spatial filter is obtained if $\Gamma_N(f)$ is replaced by $\Gamma(f)$, the whole noisefield CSDM. The adaptive beamformer is obtained by using an estimate $\hat{\Gamma}(f)$ of $\Gamma(f)$ instead of $\Gamma_N(f)$ in (8). The same filter is obtained when looking for the filter which minimizes the output spectral density with the constraint that a signal in the look direction be undistorted (minimum variance distortionless response).
Resolving power of a beamformer may be defined by using a geometrical criterion based on its response to two equal level sources : they are said to be just separated when the second derivative of the beamformer response is equal to zero at the mid-point direction : $\underline{\theta}_0 = 1/2\ (\underline{\theta}_1 + \underline{\theta}_2)$ where $\underline{\theta}_1$ and $\underline{\theta}_2$ are the directions of the two sources.

In fact, resolving power is a statistical quantity : for a given situation (geometry, signal to noise ratios), the criterion may be satisfied for a realization, but not for another one. In other words, the resolving power should be defined by the probability that two sources be resolved for a given source separation.
Nevertheless, a good characterization of resolving power is obtained by applying the criterion to the response in which the estimated quantities are replaced by their expected values.
For an array of K sensors, and with a background noise independant between the sensors, the gain in resolving power of adaptive beamforming on conventional beamforming is given by :

$$GRP_A = 0.85 \left[K \frac{\gamma(f)}{\sigma(f)}\right]^{1/4} \tag{9}$$

where $\gamma(f)$ is the spectral density of the two sources (which have equal levels) and $\sigma(f)$ the one of the background noise. Thus the resolving power of adaptive beamforming varies as the power 1/4 of the signal to noise ratio. Thus the asymptotic resolving power of adaptive beamforming is bounded by a value which increases with the signal to noise ratio.
Simulations have been carried out to verify this expression with a linear array of 9 sensors equispaced by half a wavelength. The gain in resolving power of adaptive beamforming on conventional beamforming is shown on fig. 3 versus signal to noise ratio $\gamma(f)/\sigma(f)$ for both theoretical expression (solid line) and simulation results (dotted line) : the agreement is good.

3.2 High resolution methods

High resolution methods has been extensively studied during the past decade. Their interest comes from their asymptotic resolving power which goes to infinity.
They are based on a more complete modeling of the noise-field than the one used by conventional and adaptive beamforming. The model for the sources is the same, but a model for the background noise spatial coherence is added. Its cross-spectral density matrix is written :

$$\Gamma_b(f) = \sigma(f)\ J(f)$$

where J(f), the spatial coherence matrix, is known. Without loss of generality, it can be supposed equal to the identity matrix I. The noise field is supposed to contain P sources, with P<K, statistically independent from the background noise. Thus the cross-spectral density matrix of the received signals is equal to :

$$\Gamma(f) = \mathcal{D}(f)\ \mathcal{C}(f)\ \mathcal{D}^+(f) + \sigma(f)\ I = \Gamma_s(f) + \sigma(f)\ I \tag{10}$$

where $\mathcal{D}(f)$ is a (KxP) matrix composed by the position vectors $\underline{d}(f,\underline{\theta}_i)$ of the sources and $\mathcal{C}(f)$ the (PxP) cross-spectral density matrix of the source signals which is supposed of rank P : if the sources are independent, $\Gamma(f)$ becomes equal to :

$$\Gamma(f) = \sum_{i=1}^{P} \gamma_i(f)\, \underline{d}(f,\underline{\theta}_i)\, \underline{d}^+(f,\underline{\theta}_i) + \sigma(f)\, I \tag{11}$$

High resolution method are based on the properties of the eigen decomposition of $\Gamma(f)$: the observation space (dimension K) can be divided in two subspaces :

- the signal subspace spanned by the P eigenvectors $\underline{V}_i(f)$, $i \in [1,P]$ related to the P maximum eigenvalues of $\Gamma(f)$: $\lambda_i(f) = \lambda_{si}(f) + \sigma(f)$ which are the non-zero eigenvalues of $\Gamma_s(f)$ which has rank P. It is the subspace spanned by the P position vectors of the sources $\underline{d}(f,\underline{\theta}_i)$. An important relation is :

$$\Gamma_s(f) = \sum_{i=1}^{P} [\lambda_i(f) - \sigma(f)]\, \underline{V}_i(f)\, \underline{V}^+_i(f)$$

- the noise subspace spanned by the (K-P) eigenvectors $\underline{V}_i(f)$, $i \in [P+1,K]$ related to the (K-P) minimum eigenvalues of $\Gamma(f)$ all equal to $\sigma(f)$.

This subspace is orthogonal to the signal subspace, thus a key relation is : $\underline{V}^+_i(f)\, \underline{d}(f,\underline{\theta}_j) = 0$, $\underline{V}_i(f) \in [P+1,K]$. For that reason the noise subspace is also called orthogonal subspace.

These properties are asymptotic ones. But, it can be shown [3] that the same relations exists between the maximum likelihood estimated of $\sigma(f)$, $\gamma_i(f)$ and $\underline{d}(f,\underline{\theta}_i)$ $(i\in[1,P])$ and the eigenspace of the estimate of $\Gamma(f)$ (relation (7)).

In order to estimate the source positions, either subspace can be used. However, no optimum way can be defined, and a number of "ad hoc" methods has been proposed which lead to very similar results [4 to 16]: but they have all to use a source position vector model.

High resolution methods has the unique property to have a resolving power which is no longer limited by the signal to noise ratio of the sources as with previous array processing techniques : it increases theoretically with the observation time up to infinity, so that two sources can be separated, how close and how weak they may be. An approximation of the resolving power of orthogonal beamforming or Music, a high resolution method using the orthogonal subspace [7] [8] can be obtained. Its response is given by

$$G(f,\underline{\theta}) = \{\underline{d}^+(f,\underline{\theta}) \left[\sum_{i=P+1}^{K} \underline{V}_i(f)\, \underline{V}^+_i(f) \right] \underline{d}(f,\underline{\theta})\}^{-1} \tag{13}$$

where $\underline{V}_i(f)$ are the eigenvectors of $\widehat{\Gamma}(f)$ related to the (K-P) minimum eigenvalues. For an array of K sensors and two equal power sources, the gain in resolving power on conventional beamforming is given by :

$$GRP_N = 0.625 \left[N \frac{\gamma(f)}{\sigma(f)} \right]^{1/4} \qquad (14)$$

As for adaptive beamforming, the resolving power of orthogonal beamforming increases as the power 1/4 of the signal to noise ratio of the sources, but also with the power 1/4 of N, that is to say of the observation time : it is the only array processing technique which has this property. It must be noticed that the Cramer-Rao lower bound predicts a resolving power which varies as $\sqrt{N}$. Thus it is theoretically possible to do better than high resolution methods.

Simulations have been carried out in the same conditions as with adpative beamforming, and the gain in resolving power of orthogonal beamforming on conventional beamforming is shown on the same fig. 3. for both theoretical (solid line) and simulation results (dotted line) : the agreement is good.

The main performance limitations of high resolution method come from physical mismatches which exist between the real environment and the model which support the theory. The more refined is a model, the more powerful is the method, but the greater is its sensitivity. The methods are sensitive to the non-perfect spatial coherence of the sources, and to distortions from the models of the source wavefront shape and the back-ground noise spatial coherence.

For wavefront shape distortions, the sensitivity of adaptive beamforming increases with signal to noise ratio, while for high resolution methods, it is constant : the source direction vector model is not exactly orthogonal to the asymptotic noise subspace.

The cross-spectral density matrix of a source has theoretically a rank equal to one : a loss in spatial coherence increases the rank to K. That has two effects which can be simply described in the case of one source : the eigenvector, which is theoretically equal to the source position vector, is slightly disturbed, and the eigenvectors related to the (K-1) additional eigenvalues act as interfering sources with distorted wavefronts. These phenomenoms can be handled by perturbation analysis. Similarly, mismatch on the background noise spatial coherence adds interfering sources. Several methods have been developped in order to decrease the sensitivity to mismatches. Some of them consist in adding extra modeling parameters to be estimated. This has been proposed for wavefront shape distortions [13] [14] [17] and for back-ground noise spatial coherence [18 to 21]. Another interesting method is the implementation of the processing at the output of conventional beams which cover a limited angular sector. They can be formed with all the sensors and then they have adjacent directions, or with clusters of sensors steered in the same direction [22 to 24]. This technique is effective for wave front distortion and back-ground noise distribution mismatch. It is illustrated on fig. 4 and 5. In the two cases, a linear array of 32 uniformly distributed sensors is used.

Fig. 4 shows the asymptotic resolving power gain of orthogonal beamforming when the distribution of back-ground noise is assumed uniform whereas it is

equal to (1-α cosθ) (θ : bearing). Results are presented using the ouputs of the 32 sensors or of 4 clusters of 8 sensors. Fig. 5 shows the same quantity for wavefront phase distortions uniformy distributed overs $\pm \emptyset$: results are presented using the outputs of the 32 sensors, 8 conventional beams or 4 clusters of 8 sensors.
Another problem for high resolution methods is the estimation of the number of sources, P. Several detection tests for the determination of the number of sources have been proposed [25] [3], but they prove to be very difficult to use with real data. When using orthogonal beamforming, the number N_v of eigenvectors related to the smallest eigenvalues which are used in the response (Eq. (13)), acts like a threshold in a conventional detection process. The detection of sources is done on the orthogonal beamforming response, and if N_v is decreased, the detection probability of weak sources is increased, but the false-alarm (false spikes) probability as well : but this one can be lowered by the tracking process which is following any detection process in a sonar (see [26] for example).
Another interesting procedure which is well adapted to this problem is presented in the next section.

4. UNIFIED HIGH RESOLUTION DIRECTION FINDING AND BEAMFORMING

High resolution direction finding procedures require a prior estimate, $\hat{P}$, of the number of sources, P, which are present. Specifically, for eigenstructure based methods, the estimated dimensionality of the source subspace $\hat{P}$, is equal to the maximum number of sources which can be extracted. Likewise, $\hat{P}$ specifies the minimum order for a spatial linear prediction filter which, in turn, stipulates the number of poles and therefore the estimated number of sources which characterize the spatial spectrum shaping filter used to parametrically model the source environment [27]. If $\hat{P} < P$, then not all sources will be "captured" by the direction finding process. On the otherhand, if $\hat{P} > P$, then extraneous estimation noise is introduced which, in the case of low level sources which are near the detection threshold, degrades detection performance.
By combining the robustness of a beamforming (BF) technique, which requires no prior estimate of the number of sources P to perform both the waveform estimation and the direction finding functions, with a high resolution method (HR) procedure the advantages of both approaches are achieved. In particular, a minimum variance distortionless response (MVDR) beamformer has been combined with an eigenstructure decomposition procedure [10, 22, 28]. Moreover, a conventional time delay-and-sum beamformer has been combined with linear prediction based aperture extrapolation [29]. In these unified approaches, the estimated source subspace dimensionality $\hat{P} = D$, where D is the number of dominant, i.e. high level sources which can easily be estimated from a cursory examination of the data. The combined BF/HR process provides increased resolution of the high level sources commensurate with a HR procedure and improved detection performance of the low level sources at the BF output due to both increased dominant source resolution and reduced dimension (D<P) eigenstructure estimation noise for the HR process.

In the following, the MVDR BF and eigenstructure HR combined process is examined. Specifically, it is shown how this BF/HR process can be interpreted as a generalized eigenvector beamformer null steering operation with correspondingly simple implementation in either element or beam space. First, consider the MVDR BF filter K-vector for a K sensor array specified at frequency f given by :

$$\underline{H}(f,\underline{\theta}) = \Gamma^{-1}(f)\ \underline{d}(f,\underline{\theta})\ [\underline{d}^{+}(f,\underline{\theta})\ \Gamma^{-1}(f)\ \underline{d}(f,\underline{\theta})]^{-1} \tag{15}$$

$\Gamma(f)$ can be written :

$$\Gamma(f) = \sum_{i=1}^{P} \lambda_i(f)\underline{V}_i(f)\underline{V}_i^{+}(f) + \sigma(f)\ I \tag{16}$$

$$= Mp\ Up\ Mp^{+} + I_K \tag{17}$$

The K-by-P eigenvector modal matrix Mp has the source subspace eigenvector $\underline{V}_i(f)$ as the ith column, $Mp^{+}\ Mp = Ip$, $Up = \text{diag}\ [\lambda_1(f), \lambda_2(f), \ldots, \lambda_S(f)]$. Now assume an observable enhanced dominant mode CSDM model

$$\Gamma(e,D) = e \sum_{i=1}^{D} \lambda_i(f)\ \underline{V}_i(f)\ \underline{V}_i^{+}(f) + I_K \tag{18}$$

$$= M_D\ U_D\ M_D^{+} + I_K \tag{19}$$

where $P > D$ and e is a scalar referred to as the enhancement factor, $M_D^{+}\ M_D = I_D$, $U_D = \text{diag}\ [e\ \lambda_1(f), e\ \lambda_2(f), \ldots, e\ \lambda_D(f)]$.
Next, from the Enhanced Dominant Mode (EDM) MVDR beamformer filter vector

$$\underline{H}(e,D) = \Gamma^{-1}(e,D)\ \underline{d}(f,\underline{\theta})\ [\underline{d}^{+}(f,\underline{\theta})\ \Gamma^{-1}(e,D)\ \underline{d}(f,\underline{\theta})]^{-1} \tag{20}$$

by substituting Eq. (18) into Eq. (15). Using the EDM form for the CSDM inverse

$$\Gamma^{-1}(e,D) = I_K - M_D\ (I_D + U_D^{-1})^{-1} M_D^{+} \tag{21}$$

$$= I_K - \sum_{i=1}^{D} [e\ \lambda_i(f)/(1+e\ \lambda_i(f))]\ \underline{V}_i(f)\ \underline{V}_i^{+}(f) \tag{22}$$

$$= I_K - \sum_{i=1}^{D} b_i(e)\ \underline{V}_i(f)\ \underline{V}_i^{+}(f) \tag{23}$$

to obtain

$$\underline{H}(e,D,\underline{\theta}) = (I_K - \sum_{i=1}^{D} b_i(e)\ \underline{V}_i(f)\ \underline{V}_i^{+}(f)\ \underline{d}(f,\underline{\theta})/$$

$$(K- \sum_{i=1}^{D} b_i(e)\,|\underline{V}_i^+(f)\ \underline{d}(f,\underline{\theta})|^2 \tag{24}$$

The beamformer response pattern is

$$b_{EMV}(e,D,\underline{\theta}) = \underline{H}^+(e,D,\underline{\theta})\ \Gamma(f)\ \underline{H}(e,D,\underline{\theta}) \tag{25}$$

$$= \underline{d}^+(f,\underline{\theta})\ \Gamma^{-1}(e,D)\ \Gamma(f)\ \Gamma^{-1}(e,D)\ \underline{d}(f,\underline{\theta}) / [\underline{d}^+(f,\underline{\theta})\ \Gamma^{-1}(e,D)\ \underline{d}(f,\underline{\theta})]^2 \tag{26}$$

For e approaching infinity, $b_i(e)$ in Eq. (23) approaches zero such that the denominator term in the beam response of Eq. (26) becomes

$$\lim_{e \to \infty} \underline{d}^+(f,\underline{\theta})\ \Gamma^{-1}(e,D)\ \underline{d}(f,\underline{\theta}) = \sum_{i=D+1}^{K} |\underline{d}^+(f,\underline{\theta})\ \underline{V}_i(f)|^2 \tag{27}$$

Eq. (27) is identical with the denominator of the orthogonal beamforming (or Music). Thus, the EDM MVDR beamformer gives Music-like resolution in the $\underline{\theta}$ regions near the D dominant interfering source locations and post-null steering conventional beamformer performance for the $\underline{\theta}$ regions near the low level sources. This null-steering interpretation follows from Eq. (24) wherein the output of generalized eigenbeamformers, $b_i(e)\,|\underline{V}_i^+(f)\ \underline{d}(f,\underline{\theta})|^2$ $(1 \leq i \leq D)$ for the D dominant modes, can be seen to be substracted from the output of a conventional beamformer $\underline{d}(f,\theta)$. The value of e can be varied $(e \geq 1)$ to control the depth of the null and D can be obtained from the conventional beamformer. Computation requires only the largest D eigenpairs.

A three source example is now considered. An array with N=32 sensors spaced uniformly at one third of a wavelength is modeled. Two sources with SNR_1 = 0 dB and SNR_2 = -2 dB at a sensor, are located at angles $\theta_1 = 78°$ and $\theta_2 = 84°$ and a third source with SNR_3 = -22 dB is located at $\theta_3 = 102°$. The EMD MVDR (e=1) output of Eq. (26) is shown on fig. 6 for the cases wherein D = 1,2 and 3. It is observed that the lower level source is consistently evident regardless of the dominant source subspace dimensionality D.

5. WIDEBAND HIGH RESOLUTION ARRAY PROCESSING

High resolution methods have been developped in a "narrow band" context : their asymptotic theory is valid only at a pure frequency. The question of how to handle wideband signals has been investigated by several authors.

Wideband high resolution methods has been first developped starting from the space time correlation matrix of the received signals [30]. But the result of this method turns to be an incohent integration of the output of

the different "narrow band" processing forming the whole bandwidth. This processing, which is equivalent to those used in other array processing methods, does not improve the resolving power : in fact, it decreases the resolving power compared to the narrowband solution using the highest frequency of the bandwidth, but it increases the stability and thus the detection of weak sources.
Other methods attempt to transform the CSDM estimate at each frequency in order to make the direction vector of each source independant of the frequency.

Therefore, the transformed cross-spectral density matrix estimates can be integrated, which results in a matrix whose eigenspace is an improved estimate of the actual new eigenspace which is independant of the frequency.
The problem is that this method, to be perfect, needs to know the directions of the sources.
If good estimates $\underline{\theta}_i$ of the source directions can be obtained, it is possible to define a so-called "focussing" matrix T(f) at each frequency, such that the P direction vectors for the P sources at the frequency f are transformed into the P direction vectors at a reference frequency f_0 [31] :

$$T(f)\ \underline{d}(f,\underline{\theta}_i) = \underline{d}(f_0,\underline{\theta}_i) \qquad i \in [1,P]$$

An improved version has been proposed in [32] : the focussing matrix shifts the signal-subspace estimate at the frequency f_0 (a refinement extends the signal-subspace by adding two symetrical direction vectors to each direction vector estimate).
Another way to obtain an approximate "focussing" of the eigenspace for different frequencies is to steer the signals received on the sensors for each look-direction successively [31] [33]. Thus the direction vector for a source in the look direction is independant of the frequency : but this is less and less true for the sources whose directions are more and more far from the steered direction. A high level source far from the steered direction is, after the integration of the cross-spectral density matrix at the different frequencies, transformed into a kind of correlated background noise (the corresponding cross-spectral matrix is full rank instead of rank one, but is not an identity matrix) which is not in agreement with the high resolution methods assumptions. Better results can be obtained by applying this technique after conventional beam-forming which limits the spatial sector taken into account around the steered direction.Another method has been proposed [34] which does not need any a priori knowledge of the source positions. It assumes a linear array with equally spaced sensors. The principle is to build, by interpollation from the actual sensors, for each frequency bin a "synthetic array" with a sensor spacing equal to :

$$d_i = d_0 \frac{f_0}{f_i}$$

such that for any source direction θ the phase difference between two adjacent "synthetic sensors" is constant :

$$\phi(f_i) = 2\pi f_i \frac{d_i}{c} \sin\theta = 2\pi d_0 \frac{f_0}{c} \sin\theta \qquad (28)$$

Thus, the direction vectors of a source at the frequency f_i on the "d_i spacing" array is identical to its direction vector at the frequency f_0 on the "d_0 spacing" array, and this independently of the source direction. Therefore, using the appropriate array for each frequency bin of the desired bandwidth B, we obtain cross-spectral density matrices which can be directly integrated.
An important feature for wideband high resolution methods is that, for a signal with a flat spectrum, the resolving power increases compared to the resolving power at the highest frequency of the bandwidth, but with a limit due to the fact that the widening of the bandwidth tends to increase the resolving power due to the integration gain, but the extension of the bandwidth towards low frequencies tends to decrease the resolving power due to the shortening of the array aperture in wave-length.

6. CONCLUSION

High resolution methods have been an important field of investigation for the signal processing community, but the ultimate censor remains their performance in the real world. Some experimentations have been conducted which show the improvement afforded by these methods. Some results obtained with a towed array are show on fig. 7 in a third octave bandwidth for the conventional beamformer, the incoherent orthogonal beamformer and the coherent orthogonal beamformer.

REFERENCES

[1] L. KOPP, D. THUBERT "Bornes de Cramer-Rao en traitement d'antenne" Traitement du Signal, Vol. 3, n° 3, 1986, pp. 111-125 et Vol. 4, n° 1, 1987, pp. 57-71

[2] P.M. SCHULTHEISS, K. WAGNER "Active and passive localisation : similarities and differences" Nato Asi, Kingston, Canada, 1988 ; in "Underwater Acoustic Data Processing" Y.T. CHAN ed., Kluwer Academic Pub., pp. 215-232

[3] G. BIENVENU, L. KOPP "Optimality of high resolution array processing using the eigensystem approach" IEEE Trans. on ASSP, Vol. ASSP-SI, n° 5, Oct. 1983

[4] N.L. OWSLEY "Signal set extraction", in Aspects of Signal Processing ed. G.Tacconi, Reidel, Boston, 1977

[5] S.S. REDDI "Multiple source location : a digital approach" IEEE Trans. Aer. Elect. Systems, Vol. AES-15, n° 1, Jan. 1979, pp. 95-105

[6] G. BIENVENU "Influence of the spatial coherence of the background noise on high resolution passive methods" Proceed. ICASSP 79, Washington DC, 2-4 April 1979, pp. 306-309

[7] G. BIENVENU, L. KOPP "Principe de la goniométrie passive adaptative" 7è Colloque GRETSI, NICE, 28 mai-2 juin 1979, pp. 106/1-106/10

[8] R.O. SCHMIDT "Multiple emitter location and signal parameter estimation" Proc. RADC Spectrum Estimation Workshop, Oct. 1979 pp. 243-258

[9] R. KUMARESAN, D.W. Tufts "Data-adaptive principal component signal processing" Proceed. 19th IEEE Int. Conf. on Decision and Control, Albuquerque, NM, Dec. 10-12 1980, pp. 949-954

[10] N.L. OWSLEY, J.F. LAW "Dominant mode power spectrum estimation" Proceed. ICASSP 82, Paris (France), May 3-5 1982, pp. 775-778

[11] D.W. TUFTS, R. KUMARESAN "Estimation of frequencies of multiple sinusoïds : making linear prediction perform like maximum likelihood" Proceed. IEEE Vol. 70, n° 9, Sept. 1982, pp. 975-989

[12] M. WAX, T.J. SHAN, T. KAILATH "Covariance eigenstructure approach to 2-D harmonic retrievial" Proceed. ICASSP 83, Boston MA, 14-16 April 1983, pp. 891-894

[13] H. MERMOZ "Imagerie, corrélation et modèles" Annales des Télécommunications t. 31, n° 1-2, Janv.-Fév. 1976, pp. 17-36

[14] G. VEZZOSI "Wavefront separation by rectification of the Choleski factor of the cross-spectral matrix" Proceed. Intern. Conf. on Spectral Analysis and its use in Underwater Acoustics, Imperial College, London, 29-30 April 1982, pp. 22-1 - 22-8

[15] R. KUMARESAN, D. TUFTS "Estimating the Angles of Arrival of Multiple Plane Waves" IEEE Trans. on Aerospace and Electronic Systems, Vol. 19, n° 1, January 1983, pp. 134-139

[16] J.L. LACOUME, B. BOUTHEMY, P. GLANGEAUD, C. LATOMBE, A. SILVENT "Caracterisation par analyse interspectrale du champ d'ondes reçu sur un réseau de capteurs" 7è Colloque GRETSI, Nice, 28 mai-2 juin 1979, pp. 80/1-80/7

[17] J. MUNIER, G. JOURDAIN, G.Y. DELISLE "A new algorithm for the identification of distorted wavefronts" Nato ASI, Kingston, Canada, 1988, in "Underwater Acoustic Data Processing", Y.T. Chan ed., Kluwer Academic Pub. pp. 87-92

[18] G. BIENVENU, L. KOPP "Adaptativity to background noise spatial coherence for high resolution passive methods" Proceed. ICASSP 80, Denver CO, April 9-11 1980 pp. 307-310

[19] J.P. LECADRE, P.Y. ARQUES "Estimation for aray processing of spatial noise correlations in the presence of sources" Nato ASI, Kingston, Canada, 1988, in "Underwater Acoustic Data Processing" TT. Chan Ed. Kluwer Academic Pub. pp. 357-365

[20] R. ROY, A. PAULRAJ, T. KAILATH "Estimation of Signal Parameter via Rotational Invariance Techniques", Proceedings of SPIE, Aug. 1986, San Diego, pp. 94-101

[21] J.P. REILLY, K.M. WONG "Direction of arrival estimation in the presence of noise with unknown arbitrary covariances matrices" ICASSP 89, Glasgow, 23-26 May 1989

[22] N.L. OWSLEY "Signal subspace based minimum-variance spatial array processing" Proc. Asilomar Conf. on Circ. Sys. and Com. Nov. 1985, pp. 94-97

[23] G. BIENVENU, L. KOPP "Decreasing high resolution method sensitivity by conventional beamformer preprocessing" Proceed. ICASSP 84, San Diego, CA, March 19-21 1984, pp. 33/2/1-4

[24] P. FORSTER, G. VEZZOSI "Application of spheroïdal sequences to array processing" Proceed. ICASSP 87, Dallas, 6-9 April 1987, pp. 2268-2271

[25] M. WAX, T. KAILATH "Determining the number of signals by information theoritic criteria" ICASSP 84, San Diego

[26] T.E. FORTMANN, Y. BAR-SHALOM, M. SCHEFFE "Sonar tracking of multiple targets using joint probabilistic data association" IEEE Journal Ocean. Eng., Vol. OE-8, n° 3, July 1983, pp. 173-183

[27] J. BURG "Maximum entropy spectrum analysis" 37th Annual Mtg. of Exploration Geophysicrists, Oklahoma City, 1967

[28] N.L. OWSLEY "Enhanced Minimum Variance Beamforming" Proceedings of the NATO Advanced Study Institute for Underwater Acoustic Signal Processing, Kingston, Ontario, Canada, July 1988

[29] R. WALKER, J. WINKLER APEX Proceedings of the NATO Advanced Study Institute for Underwater Acoustic Signal Processing, Kingston, Ontario, Canada, July 1988

[30] G. BIENVENU "Eigensystem properties of the sampled space-time correlation matrix" ICASSP-83, Boston, April 1983, pp. 332-335

[31] H. WANG, M. KAVEH "Coherent signal-subspace processing for the detection and estimation of angles of arrival of multiple wideband sources"
IEEE Trans. Acoust. Spee. Sign. Proc., Vol. ASSP-33, Aug. 1985 pp. 823-831

[32] H. HUNG, M. KAVEH "Focussing matrices for coherent signal subspace processing"
IEEE Trans. Acoust. Spee. Sign. Proc.,
Vol. ASSP-36, Aug. 1988 pp. 1272-1281

[33] D.N. SWINGLER, R.S. WALKER, J. KROLIK
"High resolution broadband beamforming using a doubly-steered coherent signal-subspace approach" ICASSP-88 Proceedings, NEW-YORK, April 1988 pp. 2658-2661

[34] G. BIENVENU, P. FUERXER, G. VEZZOSI, L. KOPP, F. FLORIN "Coherent wideband high resolution processing for linear array" ICASSP 89, Glasgow, 23-26 May 1989

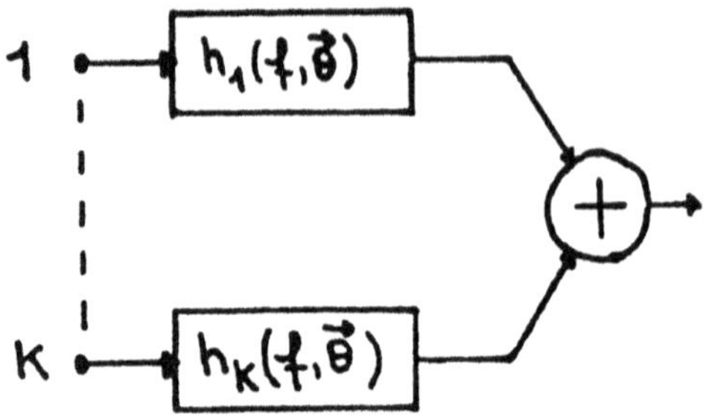

Fig. 1 Spatial processing

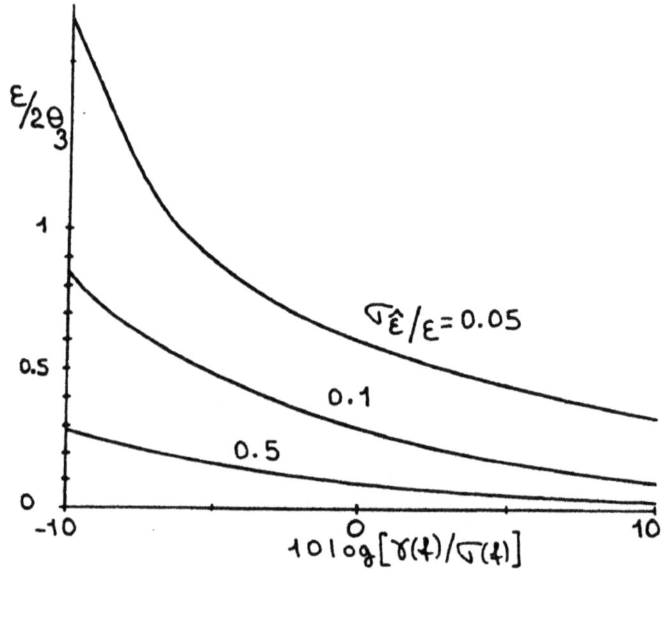

Fig. 2 Normalized source separation $\varepsilon/2\theta_3$ versus the signal to noise ratio $\gamma(f)/\sigma(f)$ for different values of $\sigma_\varepsilon/\varepsilon$. Linear array with 9 sensors equally spaced of half a wavelength

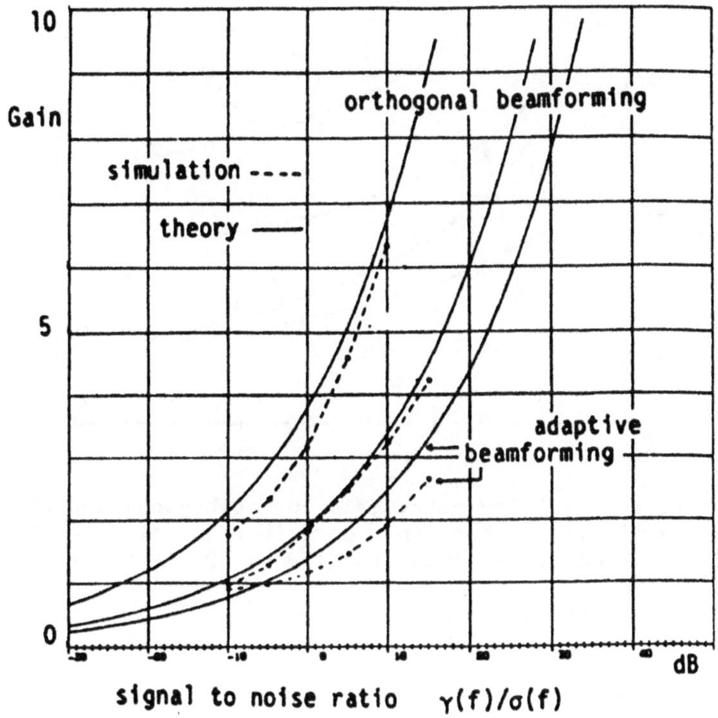

Fig. 3 Gain of orthogonal and adaptive beamforming on conventional beamforming : theory (solid lines) and simulation results (dotted lines) (same array as in Fig. 2) versus the signal to noise ratio : $\gamma(f)/\sigma(f)$

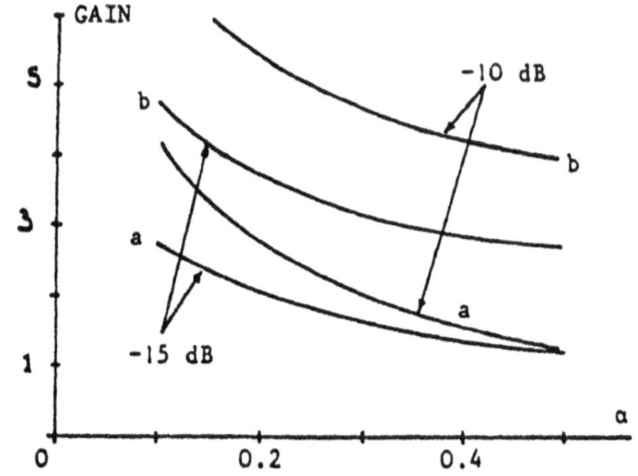

Fig. 4 Asymptotic resolving power gain of orthogonal subspace method on conventional beamforming when the distribution of background noise is assumed uniform whereas it is equal to (1-αcos 4θ) (θ : bearing)
a) output of 32 sensors
b) output of 4 clusters of 8 sensors

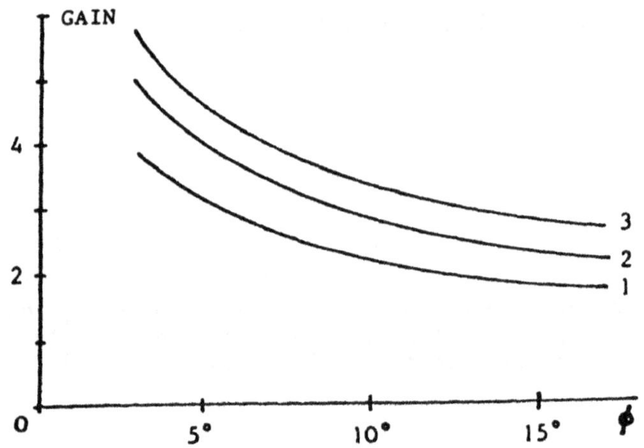

Fig. 5 Asymptotic resolving power gain of orthogonal subspace method on conventional beamforming with wavefront phase distortions uniformly distributed over ± Ø
1) 32 sensors outputs
2) 8 conventional beams outputs
3) 4 clusters of 8 sensors outputs

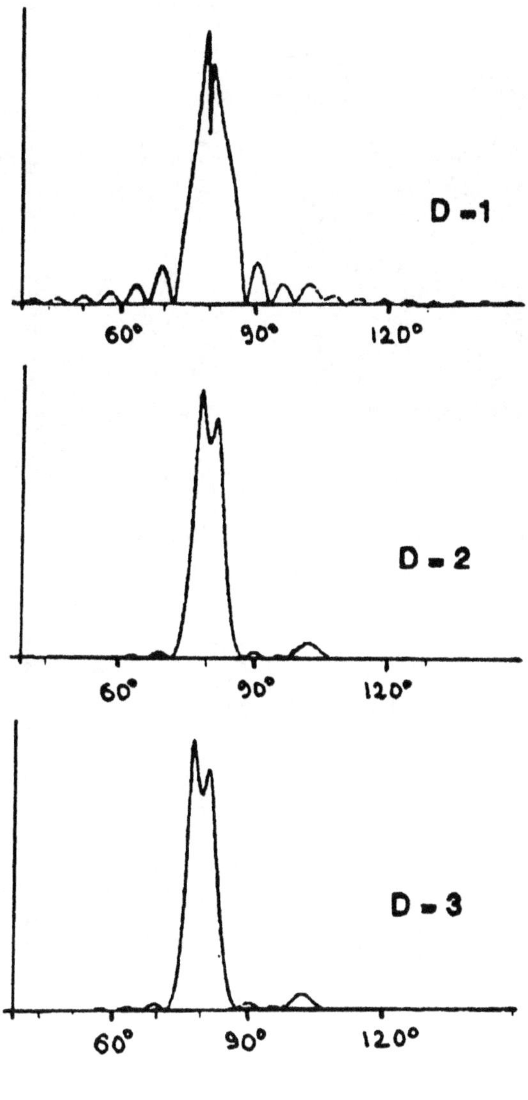

Fig. 6 Dominant mode minimum variance beamformer angular response with D=1,2 and 3 showing ability to detect source at 102° in all cases

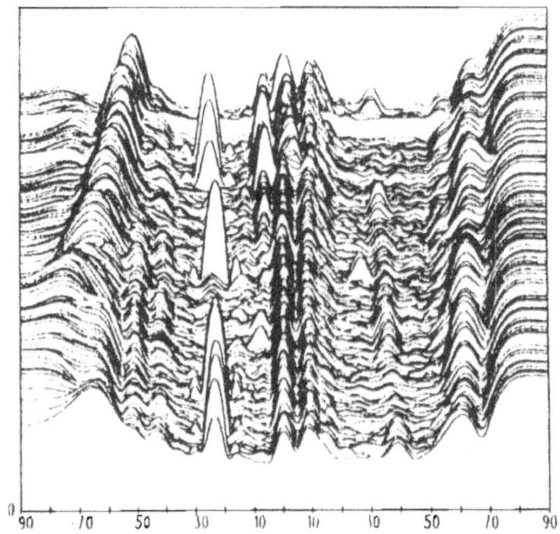

conventional beamforming

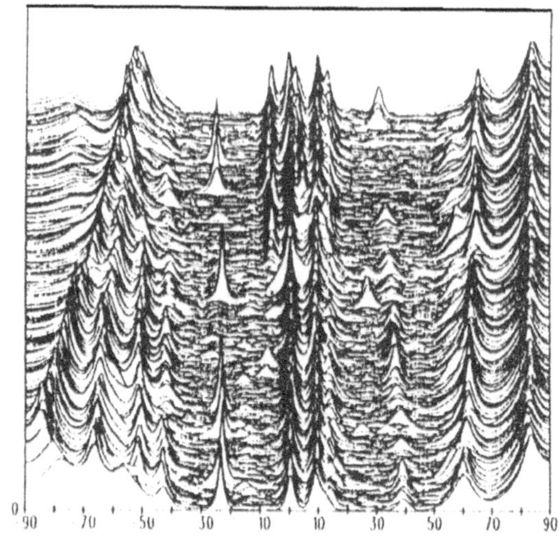

incoherent orthogonal beamforming

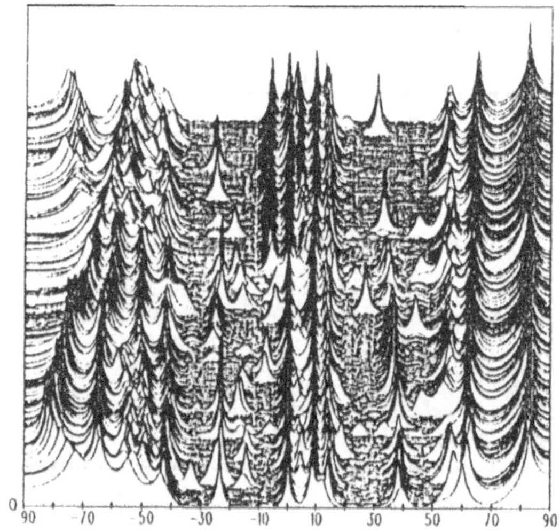

coherent orthogonal beamforming

Fig. 7 Comparison on real data of wideband conventional beamforming, incoherent orthogonal beamforming and coherent orthogonal beamforming

CHAPTER 2

EXTENSION OF HIGH-RESOLUTION METHODS TO REALISTIC CASES

J.-P. LE CADRE *
G.E.R.D.S.M., Le Brusc (France)

P. RAVAZZOLA **
G.E.R.D.S.M., Le Brusc (France)

*: now, with IRISA, Rennes (France)
**: now, with AERO, Paris (France)

PRACTICAL USE OF PARAMETRIC METHODS FOR ARRAY PROCESSING

J-P. Le Cadre P. Ravazzola

1 - INTRODUCTION

Array processing consists mainly in describing the source field received by an array of sensors. In practical situations, however, the sensor number is bounded and, therefore, the performances of classical methods (beamforming) encounter severe limitations. Actually, the received field is two dimensional (time and space), leading to the following fundamental idea: exchange observation space (limited by array length) against time (temporal integration). Assuming a perfect temporal stationarity, the problem (source bearing estimation) reduces to a purely deterministic one. Adding the hypothesis of noise whiteness, a solution can be obtained by means of elementary algebra. These considerations can be found in the seminal works of G. Bienvenu [1], [2] and have lead to high resolution (HR) methods, and finally to a tremendous amount of papers.The statistical performances of HR methods have been studied in details [3]. Nevertheless, all these methods suffer from *a priori* hypotheses (noise whiteness essentially). The consequences of noise model misadjustment may be dramatic for their practical use, therefore it is necessary to improve their robustness. This may be achieved by several means which can be essentially divided in two parts: the first one consists in estimating the noise model in order to whiten the data, while the second consists in using more fundamentally (than HR methods) the plane wave hypothesis in order to extract sources. Some aspects of these two approaches will be detailed later.

The first one consists usually (e.g. in spectral analysis) in an ARMA modelling of the data; roughly speaking, peaks correspond to the AR part whereas colored noise is associated to the MA part. Unfortunately, this approach is non valid for array processing, especially in the underwater acoustic context, where the number of sources, sensors, noise parameters make this method for the least illusory. In fact the problem's context needs specific methods, but the basic idea consists in estimating the "smooth" part of the spatial density of the received field without consideration of the source parameters.

While the previous way has been scarcely considered (for that context), the second one has motivated several approaches which can be divided in the following classes:

1 - HR methods behind beamforming [4] [5] [6]
2 - Use of special properties of the noise field [7] [8]
3 - Use of state space models [9] [10] [11]

Note that the basic idea consists in extracting the spatial frequencies (sources) without consideration of the noise field, thus it is opposite to the previous one.Class 1 and class 3 correspond to the same idea: try to take into account the plane wave hypothesis in order to extract sources from noise. For these two approaches no noise model is required. On the opposite, class 2 uses also special properties of the noise field e.g: invariance by array rotation [7] or by symmetry [8].

In fact, the problem is mainly relevant to measure theory. The received field is the sum of a field with discrete measure associated to point sources) and a continuous measure (associated to the noise). Despite this measure formulation, it is mainly considered as an algebraic problem that is far too restrictive and amounts to occulting the true nature of the problem. Following these ideas, the two approaches will be detailed along this paper; but let us now present briefly the more important drawbacks induced by noise correlations.

2 - CONSEQUENCES OF SPATIAL NOISE CORRELATIONS FOR ARRAY PROCESSING [12], [13]

Considering a random process X (t, M) with the hypothesis of spatio-temporal stationarity whose correlation function is defined by: $R(\tau, r) = \mathbb{E}\left[X(t, M) \cdot \overline{X}(t - \tau, M - r)\right]$

its spatio-temporal density is given by the classical formula [14]:

$$P(f, k) = \int_{\mathbb{R}^2} R(t, r) \cdot \exp\left(-2i\pi\,(f.\tau + k.r)\right) d\tau \cdot dr \tag{2.1}$$

(for a linear array, r becomes r and k becomes the scalar k; $k = \frac{1}{\lambda}\cos\theta$ with λ: wavelength and θ: bearing).

By partial integration, P (f, k) may be expressed as: $P(f, k) = \int_{\mathbb{R}} \gamma(f, r)\ \exp(-2i\pi\, k.r)\, dr$

($\gamma(f, r)$: cross-spectral density)

itself replaced by an approximation, i.e.: $\widehat{P}_{ns}(f, k) = \sum_{j=-n_s+1}^{n_s-1} \widehat{\gamma}(f, jd) \cdot \exp(-2i\pi k.jd)$ (2.2)

(n_s: sensor number ; d: array sample)

Note that P (f, k) (or its approximation $\widehat{P}_{ns}$ (f, k)) is the basic tool for spatial analysis. Usually, one considers physical noise model [12] (surface, isotropic, impulsive noise...) but the mapping between the noise model and its corresponding spatial density is ill-conditioned [13]. This advocates AR(MA) noise modelling; furthermore it is possible to describe any physical noise model by a low order AR(MA) model [13].

Considering now frequent hypothesis where the correlation length of the noise is short, the following results are available.

Let B be the noise CSM at a given frequency and P_B (f, k) its associated spatio-temporal density then its eigenvalues can be approximated by the samples of P_B (f, k) , (λ_m eigenvalue: $\lambda_m = P_B\left(f, \frac{m}{2\,n_s}\right)$) , and the corresponding eigenvectors by steering vectors $D_{m/2\,n_s}$ [12], [13], [14].

This approximation results simply from the approximation of the band-Toeplitz matrix B by a circulant matrix C whose eigensystem is given above [13]. Furthermore it is valid for low dimension of B (for example: dim (B) = 30, correlation length = 5, give excellent approximation [13]). The consequences of that approximation for array processing are numerous and especially important for HR methods, let us now detail and illustrate them.

2.1 - SPURIOUS SOURCES

The spurious sources can be explained by the shape of P_B (f, k); indeed the eigenvectors associated to its peaks are suitably approximated by steering vectors. Consequently spurious sources are induced by the peaks of spatial noise density.

This fact is illustrated by fig. 1 which represents the values of the projection function f(θ) for the MUSIC method when the received signal is the sum of the contributions of three sources and a correlated noise. For this example, the peak of the noise density is on the array broadside. A direct application of the MUSIC method (white noise hypothesis) gives poor results (fig. 1) for angular separation of the two sources at 35 and 45 deg., furthermore spurious sources can be seen in the vicinity of the array broadside.

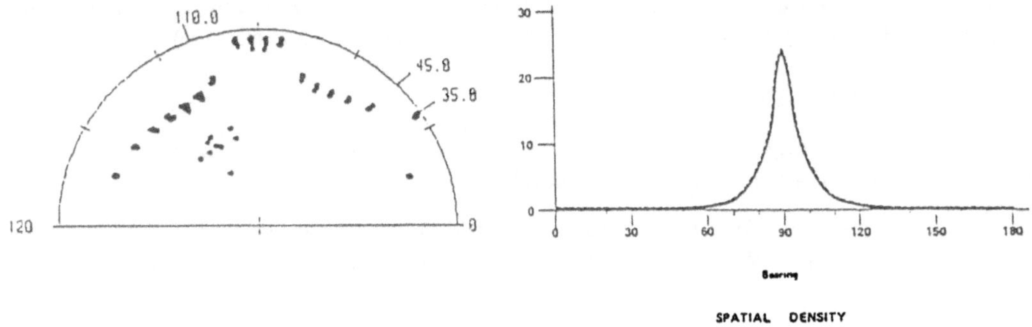

Fig. 1 : Poles of the projection function f(θ) and the corresponding noise spatial density

2.2 - NON DETECTION OF WEAK SOURCES

As it has been seen previously, a compromise between spurious sources and detection of weak sources must be determined in the presence of colored noise.

Another spurious phenomenon is illustrated by fig. 2. A weak source is simulated at a bearing corresponding to a hole of the noise spatial density. Let us precise, first, the definition of the poles of the MUSIC method which will be the basic tool.

Let $g(\theta) = D_\theta^* \cdot \Pi_B \cdot D_\theta$ be the projection function of D_θ on the noise subspace [2], then $g(\theta)$ can also be written as a function of z, i.e.: $g(\theta) = Q(z)$ with $z = \exp(-2\, i\pi\, f.d.\cos\theta / c)$

the polynomial Q (z) is positive and its roots are conjugated w.r.t. the unit circle, therefore it can be factored as:

$$Q(z) = P(z) \cdot P^*(1/z) \tag{2.3}$$

We shall call "poles" of the MUSIC method the roots of Q (z) inside the unit circle.

Then, one can see on fig. 2 that the pole associated to the weak source is attracted by the hole of the noise spatial density.

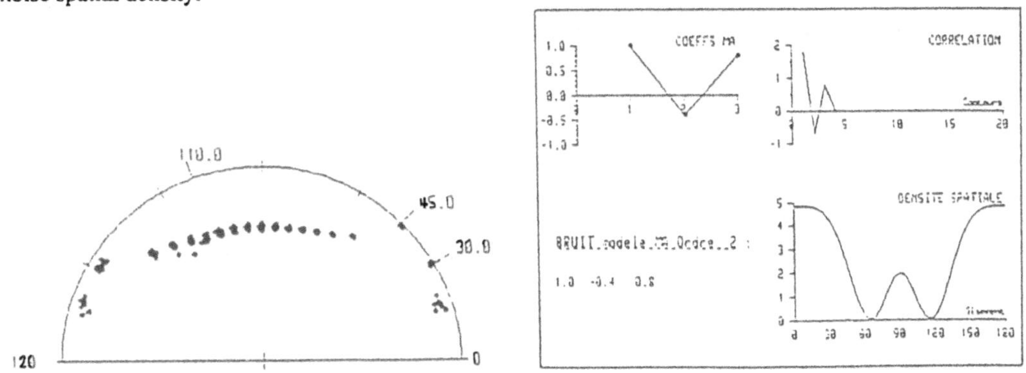

Fig. 2 : Non-detection of weak sources, attraction by the holes of the noise spatial density

This effect is verified on numerous simulations and this behavior is quite general for the MUSIC method in the presence of a colored noise, even it seems very surprising since the local signal to noise ratio is maximized for precisely that bearing. This fact may be explained by perturbation analysis and is fundamentally induced by the method itself (decomposition in source and noise subspace).

2.3 - OTHER EFFECTS

Noise coloration may also induce bias for bearing estimation and loss for angular resolution (see e.g. fig. 1) but these effects are not specific to HR methods.

As it has been shown HR methods encounter severe drawbacks due to noise correlations; estimation of a suitable noise model is of a great practical importance in these cases.

3 - ESTIMATION OF THE ADDITIVE NOISE MODELS

3.1 - GENERAL PRINCIPLES

At a given frequency (after DFT for example) the problem of separation in source and noise parts reduces to the following equation: $R = S + B$

with
$$\begin{cases} R\text{: CSM of the sensor outputs} \\ S\text{: source CSM} \\ B\text{: noise CSM} \end{cases} \tag{3.1}$$

As presented above, that problem is ill-posed and has no meaning without additional hypotheses which are:
$$\begin{cases} H_1\text{: spatial coherence [15] of sources} \\ H_2\text{: the source number is bounded} \end{cases}$$

These two hypotheses are fundamental in order to solve the separation problem. They mean that point sources cannot be described by the noise model. Generally, a separate estimation of the noise model (or B) seems preferable to a simultaneous estimation of noise and source parts.

All the methods for noise model estimation, that will be presented, obey to the following general scheme: define a barrier functional forbidding description of sources by the noise model. The estimation of the noise model will be generally achieved by means of iterative optimization methods (gradient's like) only w.r.t. noise parameters.

For the relative entropy functional (REF) method (see § 3.2) the barrier functional is the functional Log det (R - B) whereas for the likelihood method it is the likelihood functional expressed relatively to the eigenvalues of the whitened observation matrix. The two-methods are essentially different, however, since the first one (REF method) seeks the more random noise model for given observed data (the barrier functional is explicit), while the second (Maximum likelihood) uses mainly the properties of conditional likelihood as an implicit barrier functional. Obviously, it is possible, using the general scheme above, to define numerous methods for noise model estimation but the methods that will be presented develop two basic views of the problem. Let us detail them now.

3.2 - THE RELATIVE ENTROPY FUNCTIONAL APPROACH [16]

3.2.1 - Introduction

This approach will lead to define a functional depending only upon sensor CSM and additive noise CSM. This functional will be named, for the sequel, relative entropy functional (REF) and will play the central role. The REF will be maximized by iterative methods (gradient's like) relatively to the parameters defining the noise correlation between sensors.

By this way, an estimation of the noise CSM is obtained. A crucial point consists in studying the theoretical properties of that functional. The REF method appears to be well suited for array processing because the fundamental hypotheses are very often satisfied (practically) and also because the maximization of the REF involves only **noise parameters**. Therefore, for practical case (great sensor number) it is a fundamental advantage with respect to general ARMA modelling of the sensor outputs.

This method is not an additional stone in the wall of ARMA estimation methods, it uses fundamentally the hypothesis of source spatial coherence to separate the noise and source contributions. After the method has been presented, the accuracy of the estimate are theoretically studied and, then, an iterative algorithm for maximization is given together with a detailed study of convergence. Lastly, simulation results are presented in order to show the method's efficiency, its robustness.

3.2.2 - Definition of the REF

We shall consider, for the sequel, a linear array constituted by n_s equispaced sensors impinged by s sources and we shall assume that the correlation length of the noise is null beyond p sensors (p is, furthermore, little in regard to n_s).

Consider now the vector β (q x L) defined by:

$$\beta^t = \left(B_1^t, B_2^t, \ldots, B_L^t\right) \tag{3.2}$$

where the vectors B_i are (statistically) **independent** sample vectors of the noise. This last hypothesis follows from the short length correlation assumption. The parameter q is arbitrary, its major role will be to reduce the cost of computation.

Let, furthermore, X be an observation vector (sensor outputs). Now, we shall consider the concept of mutual information which have been advocated by Rao, Akaike, Arun and Kung [17], [18] in a somewhat different context (Mainly canonical correlation analysis, predictive efficiency...) and appears to be a very powerful tool.

Noting $\beta \mid X$ the linear minimum variance estimator of the zero mean random vector β from the zero mean random vector X (the orthogonal projection of the random vector β onto the Hilbert space spanned by X):

$$\beta \mid X = \mathbb{E}\left(\beta \,.\, X^*\right).\left[\mathbb{E}\left(X \,.\, X^*\right)\right]^{-1}.\, X \tag{3.3}$$

Then, a "measure" of uncertainty upon β which is not explained by X is obtained from the conditional variance of β and is equal to:

$$I\left(\beta,\ X\right) = \text{Log det}\left[\text{covar}\left(\beta - \beta \mid X\right)\right] \tag{3.4}$$

In fact, the observation vector X is the sum of a source part (S) and a noise part (B_1), let be:

$$X = S + B_1$$

Using (3.3) and (3.4), the following expression of I (β, X) is easily derived:

$$I\left(\beta,\ X\right) = \text{Log det}\left\{\begin{pmatrix} B & 0 \\ 0 & B \end{pmatrix} - \begin{pmatrix} B \\ 0 \\ 0 \end{pmatrix} . R_q^{-1} . \left(B\ 0 \ldots 0\right)\right\}$$

$$= \text{Log det}\left(R_q - B\right) + L \,.\, \text{Log det}\, B - \text{Log det}\, R_q \tag{3.5}$$

where R_q and B are two (q x q) matrices denoting the covariances matrices of X and B_i (for each i).

Roughly, the problem consists in seeking the noise structure B maximizing the uncertainty not "explained" by the statistical properties of X. This will be enlightened by the study of the properties of I (β, X).

The preceding exposure uses the exact covariance matrix of q sensors; practically R_q is replaced by an estimation $\hat{R}_q$. $\hat{R}_q$ is then obtained by orthogonal projection on the Toeplitz subspace, let be:

$$\hat{R}_q(1, i) = \frac{1}{n_s - i + 1} \cdot \sum_{j=1}^{n_s} \hat{R}(j - i + 1, j) \tag{3.6}$$

($\hat{R}$: estimated covariance matrix of the sensor outputs)

This estimation is quite satisfying, however for a little number of snapshots $\hat{R}_q$ may be non positive definite. This problem can be easily canceled, either by addition of the identity matrix multiplied by a suited factor or by estimating $\hat{R}_q$ in T^+ (the subspace of positive definite Toeplitz matrices) as defined in [17].

Now a parametrization of B is needed, it may be a linear parametrization of a Toeplitz band matrix, i.e.:

$$B = \sum_{i=1}^{p} \beta_i \cdot U_i$$

where the (q x q) matrix U_i is defined by: $U_i(k, l) = \begin{cases} 1 & \text{if } |k - l| = i - 1 \\ 0 & \text{else} \end{cases}$

It can be easily verified that matrices $\{U_i\}_{i=1}^{q}$ constitute an orthogonal basis of T_q (for the scalar product $\langle A, B\rangle = \mathrm{tr}(A \cdot B^t)$). However this parametrization does not ensure positive definiteness of B, in order to avoid such problem it is possible to parametrize B as the covariance matrix of a MA process, let be:

$$B = \left(\sum_{i=0}^{p-1} b_i \cdot Z_i\right) \cdot \left(\sum_{i=0}^{p-1} b_i \cdot Z_i\right)^t$$

where Z_i are (p, p + q) rectangular matrices given by:

$$Z_i = \begin{pmatrix} 0 \cdots 0 & 1 & 0 \cdots\cdots & 0 \\ \vdots & \ddots & \ddots & \vdots \\ 0 \cdots\cdots & 0 & 1 & 0 \cdots 0 \end{pmatrix}$$

or:

$$Z_i(k, l) = \begin{cases} 1 & \text{if } |l - k| = i \\ 0 & \text{else} \end{cases}$$

(this model being assumed minimum phase).

For the sequel, the REF will be defined as:

$$H(\beta_1, \ldots, \beta_p) = \mathrm{Log}\det(\hat{R} - B) + L.\ \mathrm{Log}\det B \tag{3.9}$$

where $B = B(\beta_1, \ldots, \beta_p)$

The role of the scalar factor L will appear more clearly later. This functional is maximized (w.r.t. the $\{\beta_i\}$) by iterative techniques, but let us consider, first, the properties of the corresponding estimates.

3.2.3 - Accuracy of noise parameters estimation

The REF is a non linear functional on the parameters $\{\beta_i\}$ of the noise covariances. Therefore it seems impossible to obtain a direct result for the estimates $\{\beta_i\}$ (maximizing H). It seems much simpler to translate the problem in terms of eigensystem.

The spatially white noise case is presented. It is not at all relevant to our problem, but it allows to obtain a simple result and to enlighten the role of the factor L. Then, the general case is considered; only the fundamental results are presented (the derivation is given in appendix A).

a) Spatially white noise case

This case is very simple and the reasoning is rather similar to the one used in the general case (appendix A).

Noise is assumed to be uncorrelated (sensor to sensor), therefore: $B = \lambda \,.\, Id$ (3.10)

(Id: identity matrix)

Consider now an eigendecomposition of $\widehat{R}_q$:

$$\widehat{R}_q = \sum_{i=1}^{q} \alpha_i \,.\, V_i \,.\, V_i^* \quad , \; V_i \perp V_j \quad (i \neq j) \,, \|V_i\| = 1 \tag{3.11}$$

$(\alpha_1 \geq \alpha_2 \geq \ldots \geq \alpha_q \geq 0)$

Then, by means of elementary algebra: $H(\lambda) = \sum_{i=1}^{q} \mathrm{Log}\left(\alpha_i - \lambda\right) + L \,.\, q \,.\, \mathrm{Log}\, \lambda$

The problem is now to determine the value of λ maximizing $H(\lambda)$.

Now:
$$\frac{\partial H}{\partial \lambda} = L \,.\, q \,.\, \frac{1}{\lambda} + \sum_{i=1}^{q} \frac{1}{\left(\lambda - \alpha_q\right) + \left(\alpha_q - \alpha_i\right)}$$

and remarking that if $\lambda = \alpha_q \,.\, (L / L + 1)$, then:

$$\frac{\partial H}{\partial \lambda} = (L+1) \,.\, \frac{q}{\alpha_q} + \sum_{i=1}^{q} \left[-\frac{1}{(L+1)} \alpha_q + \left(\alpha_q - \alpha_i\right)\right]^{-1}$$

therefore: $((\alpha_q - \alpha_i) \leq 0 \,, i = 1, 2, \ldots, q)$

$$\frac{\partial H}{\partial \lambda} \left(\alpha_q \cdot \frac{L}{L+1}\right) \geq 0 \tag{3.12}$$

Now, if λ tends towards α_q, $\partial H / \partial \lambda$ tends towards $-\infty$. H being a differentiable and concave function of λ in the interval $]0, \alpha_q[$, it can be deduced from all that precedes that H has one and only one maximum $\widehat{\lambda}$ in $]0, \alpha_q[$ and this maximum $\widehat{\lambda}$ corresponds to a value of λ satisfying the following inequalities [16]:

$$\alpha_q \,.\, (L / L + 1) \leq \widehat{\lambda} \leq \alpha_q \tag{3.13}$$

b) General case

The noise is now correlated between sensors, let B_0 be the true noise CSM ($B_0 \neq \lambda$ Id) and assume that B_0 may be described by the parametrization defining the matrices B (and regardless of the estimation errors of $\widehat{R}$); then the following properties are verified.

Property 1: *Let* $\{\lambda_i^w\}$ $(i = 1, \ldots, q)$ *the eigenvalues of the (whitened) matrix* $B_0^{-1} \,.\, B$. *Then if* $\widehat{B}$ *is the estimated matrix obtained by maximization of the REF under the two constraints (*B *and* R - B *positive definite matrices), these eigenvalues satisfy the following inequalities:*

$$(L / L + 1) \leq \widehat{\lambda}_i^w \leq 1 \qquad i = 1, 2, \ldots, q$$

$(\widehat{\lambda}_i^w$ eigenvalues of $B_0^{-1} \,.\, \widehat{B})$

The rather technical proof of property 1 is skipped in appendix A. It is stressed that the accuracy of noise parameters estimates is expressed only in terms of eigenvalues of $B_0^{-1} \cdot \widehat{B}$ and not directly in terms of parameters $\widehat{b}_i$. However these two subsets are strongly related, even if these relations are highly non linear and non explicit in the general case. The eigenvalues define a matrix except for arbitrary unitary transformations, but using particular structure of B it is possible to stand the following property.

Property 2: *There is a one to one correspondence between the parameters* $\{b_i\}$ *defining* B *and its eigenvalues.*

Indeed, consider the Jacobian matrix of partial derivatives [14]:

$$J = \begin{pmatrix} \frac{\partial}{\partial \beta_1}\lambda_1 & \cdots & \frac{\partial}{\partial \beta_1}\lambda_q \\ \vdots & & \vdots \\ \frac{\partial}{\partial \beta_p}\lambda_1 & \cdots & \frac{\partial}{\partial \beta_p}\lambda_q \end{pmatrix} \tag{3.14}$$

Then, using a classical lemma of linear of algebra (simple eigenvalues), i. e.:

$$\frac{\partial}{\partial \beta_l}\lambda_i = V_i^* \cdot \frac{\partial}{\partial \beta_l}B \cdot V_i$$

(V_i being eigenvectors of B)

one obtains:

$$J = \begin{pmatrix} tr\left(U_1 \cdot V_1 \cdot V_1^*\right) & \cdots & tr\left(U_1 \cdot V_q \cdot V_q^*\right) \\ \vdots & & \vdots \\ tr\left(U_p \cdot V_1 \cdot V_1^*\right) & \cdots & tr\left(U_p \cdot V_q \cdot V_q^*\right) \end{pmatrix} \tag{3.15}$$

(tr: meaning the trace of a matrix)

Now, the matrices $\left\{V_i \cdot V_i^*\right\}_{i=1}^{q}$ are linearly independent in M_q (C) (the space of hermitian (q x q) matrices), as the matrices $\{U_i\}_{i=1}^{p}$ form an orthogonal system, the rank of the rectangular matrix J is generally equal to p when q is given greater than p, and therefore the mapping is 1: 1.

This reasoning may be straightforwardly extended to the eigenvalues of $B_0^{-1} \cdot \widehat{B}$.

Finally, when L becomes great the eigenvalues of $B_0^{-1} \cdot \widehat{B}_L$ approach 1 (prop. 1) and the parameters defining $\widehat{B}_L$ (i.e. $\widehat{b}_j^L$) approach those defining B_0 (i.e.: b_j^0).

3.2.4 - Comments

The preceding results need some comments:

a) Prop. 1 is still valid (appendix A) when the noise model is **overdetermined**, underdetermination is more dangerous.

b) The assumptions needed for sources are:

   1 - source number inferior to q,

   2 - coherence property.

   However, the unicity of decomposition between sources and noise is understood.

c) The mismatches resulting from estimation of $\widehat{R}$ have not been considered in the proof of Prop. 1. Obviously, the introduction of an estimated $\widehat{R}$ leads to statistical results.

The last point has not been theoretically considered but simulation results will illustrate it.

The functional is the sum of two terms, the first one (i.e. Log det (R - B)) is a barrier functional avoiding description of sources (coherent part of the received signals) by the noise model; the second represents the underdetermination of the noise. This intuitive presentation of the functional will be developed later.

3.2.5 - Iterative methods for maximization of the REF H

The numerical problem consists now in determining the values of the parameters $\{\beta_i\}_{i=1}^{p}$ which maximize the REF H under the positivity constraints. After the study of convergence, a gradient method will be presented.

Relatively to the parameters $\{\beta_i\}$, the REF enjoys a nice property.

Property 3: *On the constraints domain (*B *and* R - B *positive definite) the REF is a concave functional relatively to* $\{\beta_i\}_{i=1}^{p}$.

The proof of property 3 follows from classical results of linear algebra. More precisely, using the classical lemmas of linear algebra standing that:

$$\begin{cases} \dfrac{\partial}{\partial \beta} \text{Log det B}(\beta) = \text{tr}\left(B^{-1}(\beta) \cdot \dfrac{\partial B}{\partial \beta}\right) \\[2ex] \dfrac{\partial}{\partial \beta} B^{-1}(\beta) = -B^{-1}(\beta) \cdot \dfrac{\partial B}{\partial \beta} \cdot B^{-1}(\beta) \end{cases} \tag{3.16}$$

(the lemmas are valid for any differentiable family of isomorphisms, see e.g. [18])

one obtains easily:

$$\frac{\partial^2}{\partial \beta_i\, \partial \beta_j} H = -\,\text{tr}\left[(R-B)^{-1} \cdot U_i \cdot (R-B)^{-1} \cdot U_j\right] - L\,\text{tr}\left[B^{-1} \cdot U_i \cdot B^{-1} \cdot U_j\right] \tag{3.17}$$

Let X be some vector of $\mathbb{R}^p$ ($X^t = (\beta_1, \ldots, \beta_p)$), then: $\quad X^t \cdot H_2 \cdot X \equiv \sum_{i,j} \beta_i \cdot \dfrac{\partial^2 H}{\partial \beta_i\, \partial \beta_j} \cdot \beta_j$

(definition of H_2, Hessian matrix of H)

and using (3.17) and the linearity of the trace of matrices, it follows that:

$$X^t . H_2 . X = - tr\left[(R - B)^{-1} . \left(\sum_{i=1}^{p} \beta_i . U_i\right) . (R - B)^{-1} . \left(\sum_{j=1}^{p} \beta_j . U_j\right)\right]$$
$$- L\, tr\left[B^{-1} . \left(\sum_{i=1}^{p} \beta_i U_i\right) . B^{-1} . \left(\sum_{j=1}^{p} \beta_j U_j\right)\right] \qquad (3.18)$$

The two terms tr (-) of (3.18) have the form: - tr (A.C.A.C) with:

$$A = (R - B)^{-1} \text{ or } B^{-1} \text{; and } C = \sum_{i=1}^{p} \beta_i U_i$$

The matrix A being assumed to be positive, it admits a Choleski decomposition in triangular factors, let be $A = T . T^*$, and therefore:

$$- tr [A . C. A . C] = - tr [T . T^* C . T . T^* . C] = - tr [(T^* C T) . (T^* C T)] = - \| T^* . C . T \|_F^2 \qquad (3.19)$$

($\| \ \|_F$ being the Frobenius norm of a matrix [19])

So that finally, the quadratic form $X^t . H_2 . X$ is negative for any vector X. The matrix H_2 is therefore a negative definite matrix and H is a concave functional w.r.t. the parameters $\{\beta_i\}_{i=1}^{p}$ on the constraints domain $\mathscr{C}$.. Consequently, gradient's method (with optimal step) will converge on $\mathscr{C}$. Let us now present, briefly, the gradient method.

Calculation of the gradient vector is straightforward, the i-th component of the gradient vector at iteration k is given by:

$$G_k (i) = - tr\left[(R - B_k)^{-1} . U_i\right] + L . tr\left[B_k^{-1} . U_i\right] \qquad (3.20)$$

(B_k: noise matrix at iteration k)

The general form of a gradient method is: $X_{k+1} = X_k - \rho_k G_k$

with $X_k^t = \left(\beta_1^k, \ldots, \beta_p^k\right)$ and G_k defined by (3.20) (3.21)

The scalar ρ_k is the step of the algorithm, it is worth to determine the optimal step in order to ensure convergence on the domain $\mathscr{C}$.

The matricial translation of (3.21) is: $B_{k+1} = B_k - \rho_k . D_k$,

with:

$$\begin{cases} B_k = \sum_{i=1}^{p} \beta_i^k . U_i \\ D_k = \sum_{i=1}^{p} G_k (i) . U_i \end{cases} \qquad (3.22)$$

The optimal step ρ_k can be easily determined by eigendecompositions, the corresponding algorithm is presented below (and detailed in appendix B):

Step 1: B_k and $\hat{R} - B_k$ being (by hypothesis) positive definite, decompose them in triangular factors:

$$B_k = T_k . T_k^* \ ; \ \hat{R} - B_k = S_k . S_k^* \qquad (3.23)$$

Step 2: Compute the eigenvalues $\{\lambda_i^k\}$ and $\{\mu_i^k\}$ of the two hermitian matrices:

$$S_k^{-1} . D_k . S_k^{-1\,*} \text{ and } T_k^{-1} . D_k . T_k^{-1\,*} \qquad (3.24)$$

Step 3: Then the REF becomes an explicit function of the parameter ρ, and is given by:

$$H(\rho) = \sum_{i=1}^{q} \mathrm{Log}\left(1 + \rho \, . \, \lambda_i^k\right) + L \, . \sum_{i=1}^{q} \mathrm{Log}\left(1 - \rho \, . \, \mu_i^k\right) + \mathrm{cst} \tag{3.25}$$

Furthermore (it's perhaps the most important fact) the positivity constraints $\mathscr{C}$ are translated in explicit (w.r.t. ρ) constraints; let:

$$\begin{cases} 1 + \rho \, . \, \lambda_i^k > 0 & i = 1, \ldots, q \\ 1 - \rho \, . \, \mu_i^k > 0 & i = 1, \ldots, q \end{cases} \tag{3.26}$$

Step 4: The optimal step ρ will be simply determined by maximization of $H(\rho)$ given by (3.25) under the constraints (3.26). This task is easily achieved by means of an **unidimensional** Newton method starting from 0.

The (computation) cost of steps 1, 2, 3, 4 is not negligible but it ensures convergence on $\mathscr{C}$ and, consequently, the estimates of noise correlations satisfy prop. 1 (except for errors in the estimation of $\hat{R}$). The computation time is not drastically important because in most of practical situations (large array) q may be chosen little in regard to the sensor number n_s.

3.2.6 - Some extensions

In numerous practical situations, the noise field received by the array results from a non symmetric distribution of elementary sources (of noise) relatively to the array broadside. The causes of this non symmetry are varied among them are: traffic noise, flow noise, vertical array...

For all these cases, the noise coefficients $\{\beta_i\}$ will be complex number and it necessary to extend the previous algorithm to this case; this is a direct extension detailed in [13], [16].

The case of a MA noise model is also quite similar to the previous one except for the gradient algorithm where the matrix U_i of formula (3.7) is simply replaced by the matrix V_i:

$$V_i = \left(\sum_{j=0}^{p-1} \beta_j^k \, . \, Z^j\right) . \, (Z^i)^t + (Z^i) \, . \left(\sum_{j=0}^{p-1} \beta_j^k \, . \, Z^j\right)^t \tag{3.27}$$

Other modellings can be considered (AR, ARMA, ...), but in general the REF properties are similar. An AR noise modelling can lead to fast method for REF maximization [13]. However in all the cases an important problem is underlined: what are the conditions ensuring a unique decomposition in sources and noise?

3.2.7 - Unicity of the decomposition

This section will be devoted to the choice of parameters p (noise model order) and q (computation cost). We are now concerned with fundamental following problem: unicity of the decomposition.

A sufficient and rough condition will be obtained as follows: assume the existence of two such decomposition, then:

$$R = S_1 + B_1 = S_2 + B_2$$

Thus:

$$S_1 - S_2 = B_2 - B_1 \tag{3.28}$$

Assume, furthermore, that the source number is bounded by S and the noise model order by p.

The CSM of sources S_1 corresponds to source parameters $\{q_j(1), z_j(1)\}_{j=1}^{S}$ where $q_j(1)$ and $z_j(1)$ are respectively the power and phase factor of the j-th source ($z_j = \exp(-2i\pi f_j)$; f_j spatial frequency, $i^2 = -1$), idem for S_2.

Denoting Z_j the following vector associated to source j: $Z_j^t = \left(z_j^{p+1}, \ldots, z_j^q\right)$ $j = 1, 2, \ldots, s$ then, the first row of (3.28) resumes all the available information (Toeplitz matrices) and implies the following equality:

$$\sum_{j=1}^{s} q_j(1) \cdot Z_j(1) = \sum_{j=1}^{s} q_j(2) \cdot Z_j(2) \tag{3.29}$$

Now, assume the dimension of Z_j (i.e.: q - p) superior to 2s, then using a basic result on Vandermonde determinants, the vectors $\{Z_j(1)\}_{j=1}^{s}$ and $\{Z_j(2)\}_{j=1}^{s}$ are linearly independent (all together). Therefore, with the above hypothesis, the unique solution of (3.29) is either:

$$\begin{cases} q_j(1) = 0 & j = 1, \ldots, s \\ q_j(2) = 0 & j = 1, \ldots, s \end{cases}$$

or, for each j (related to S_1), there exists k (related to S_2) such that:

$$\begin{cases} q_j(1) = q_k(2) \\ Z_j(1) = Z_k(2) \end{cases} \tag{3.30}$$

That reasoning leads to:

Property 4: *Assume that q is superior to* p + 2s, *then the decomposition in source and noise matrices is unique.*

It is necessary to emphasize that this condition is only sufficient and not necessary (in particular, the p first elements of R are not utilized) and appears to be very strong. In [13] examples are given where this condition is not satisfied while the results are quite satisfying.

3.2.8 - Spatial frequency analysis

The REF properties rely upon algebraic properties, the maximization of H needs matricial calculations (on the outputs CSM). In practice, the available observations are generally the beam outputs; therefore it will be very convenient for numerous applications to translate the expression of the REF in terms of spatial frequency. Then the multidimensional problem becomes monodimensional, the spatial density of noise is directly estimated, statistical properties of the method can be more easily investigated.

Let R (f, k) be the density (in frequency and space) of the stationary process received by the array. Using Szegö's theorem [20], one obtains (for a large array):

$$\lim_{n_s \to \infty} \frac{1}{n_s} \operatorname{Log} \det R_f = \frac{1}{2W} \cdot \int_{-W}^{W} \operatorname{Log} R(f, k)\, dk$$

with:
$$\begin{cases} W\text{: spatial bandwidth} \\ W = d/\lambda \text{ (d: intersensor distance)} \\ R_f\text{: being CSM matrix of sensor outputs at the frequency f} \\ k\text{: wavenumber} \end{cases}$$

(with the notations of [21] for instance) (3.31)

Hence for a great sensor number (n_s) the REF H takes the following form:

$$H = \int_{-W}^{W} \text{Log}\,[R\,(k) - B\,(k)]\,dk + L\,.\int_{-W}^{W} \text{Log}\,B\,(k)\,dk \qquad (3.32)$$

(the frequency is omitted for the sequel)

A moving average (MA) noise modelling seems the more appropriate, then B (k) is given by:

$$\begin{cases} B\,(k) = \sigma^2\,.\,F\,(z)\,.\,F^*\,(1/z) \\ \text{with :} \\ F\,(z) = 1 + b_1\,z^{-1} + \ldots + b_p\,z^{-p} \\ \quad z = \exp\,(2i\pi\,k\,d) \end{cases} \qquad (3.33)$$

The simpler approach consists in replacing R (k) by an estimate $\hat{R}$ (k), thus the classical (Fourier) estimate $\hat{R}$ (k) can be considered, i.e.:

$$\hat{R}\,(k) = \sum_{j=1-n_s}^{n_s-1} w_j\,.\,\hat{r}\,(f, jd)\;\exp\,(2i\,\pi\,k\,jd) \qquad ;\; i^2 = -1 \qquad (3.34)$$

For practical applications, the use of weighting factors w_j (e.g. Hamming) is necessary in order to avoid problems induced by the source sidelobes, in particular triangular weightings ensure the positivity of $\hat{R}$ (k) (which is needed).Note that classical Fourier analysis and weightings do not provide high (angular) resolution; but in fact, the aim of the method is not to estimate sources but the "smooth" part (i.e.: noise model). However if some sources are clustered, they may induce important errors for the noise model estimation; this drawback may be fortunately canceled by multifrequency analysis.

Using the expression (3.32) of H it is also possible to precise the fundamental property 1 and to give a more intuitive proof.

Property 5: *Consider that noise received by the array may be exactly modelled by a MA process* ($b_1, \ldots, b_p$)*, then for any MA modelling of the same order the coefficients of the MA process maximizing H (3.32) satisfy the following inequalities:*

$$\left|\left(b_i - \hat{b}_{i,\,L}\right).\frac{1}{b_i}\right| \leq \left(1 - \left(\frac{L}{(L+1)}\right)^{1/2}\right) \qquad i = 1, 2, \ldots, p \qquad (3.35)$$

Note that (3.35) does not take into account the estimation errors of $\hat{R}$ (k) (otherwise L could be chosen as great as possible!), it is not a statistical result. The proof of prop. 5 is given in appendix C.

Formula (3.32) gives an intuitive presentation of the method; the functional Log (R (k) - B (k)) plays the role of a barrier functional avoiding that sources be described by the noise model whereas the other term (weighted by the factor L) represents the uncertainty about the noise model. The method appears then as a means to "extract" the smooth part (noise) from the estimated spatial density; the factor L is related to the redundancy of information about noise received along the array.

The proposed method is well suited to a multifrequency estimation of the noise model. It seems realistic to assume a uniform noise modelling whatever the considered frequency [13] leading to the following form of a multifrequency REF:

$$H = \sum_{f=f_{min}}^{f_{max}} \int_{-W}^{W} \text{Log}\,[R_f\,(k) - B_f\,(k)]\,dk + L\,.\sum_{f=f_{min}}^{f_{max}} \int_{-W}^{W} \text{Log}\,B_f\,(k)\,dk \qquad (3.36)$$

The maximization procedure of H is similar to the monofrequency case but the advantages may be important; indeed, source bearings are identical whatever the considered frequency but this is not true for the spatial density of the noise [22].

Obviously it is possible to replace the basic assumption about noise model and to consider an expansion of the noise coefficients.

3.2.9 - Simulation results

The behavior of the method has been illustrated by numerous simulation results as detailed in the referenced paper [16]. Statistical effects due to estimation errors on R_q have been studied, the aim of these simulations was to stress the gap between the exact properties and the statistical results. In that spirit, the statistical effects of the choice of L are especially considered.

In the asymptotic case (R_q perfectly estimated) the fundamental property 1 (or 5) is satisfied. Prop. 1 is satisfied even for strong signal to noise ratios and the results (for estimation of noise parameters) are very slightly degraded when source powers increase. The results seems more sensitive to source bearings (w.r.t. noise spatial density) [13], effects of noise modelling overdetermination have also been considered. In all the cases prop. 1 is still satisfied. If the source number is great (w.r.t. n_s) prop. 1 is no longer valid but the results are not dramatically modified.

Replacing the matrix R by an estimate $\widehat{R}$ leads to statistical results for noise parameters estimation. Obviously, prop. 1 leads to chose L as great as possible, but this has no statistical meaning... The fundamental question is then: what are the effects of estimation, is it possible to find an "optimal" L?

The "quality" of the noise parameter estimates have been investigated by means of the eigenvalues of the matrix $B_0^{-1} \cdot \widehat{B}_L$. Due to the effects of errors in estimation of R prop. 1 no longer holds, especially for great values of L.

However the gap with asymptotic results does not remain too important (see fig. 4). We can remark, furthermore, that the variance of the eigenvalues of $B_0^{-1} \cdot \widehat{B}_L$ is much more important for great values of L as illustrated by fig. 3 and fig. 4. This fact advocates the use of a factor L given by the heuristic "rule" $L = n_s / 2p$ (which is in this case L = 3)

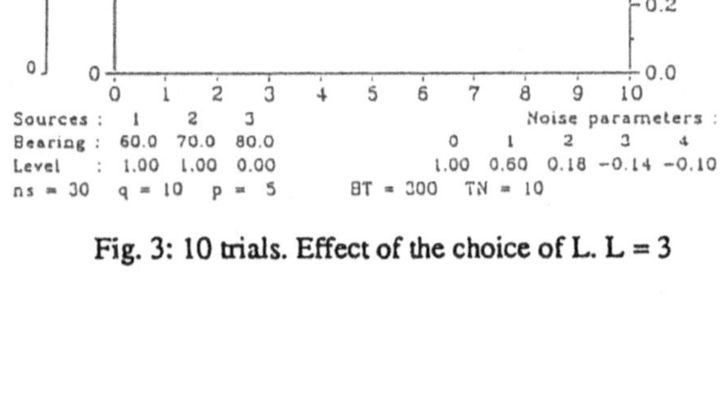

Fig. 3: 10 trials. Effect of the choice of L. L = 3

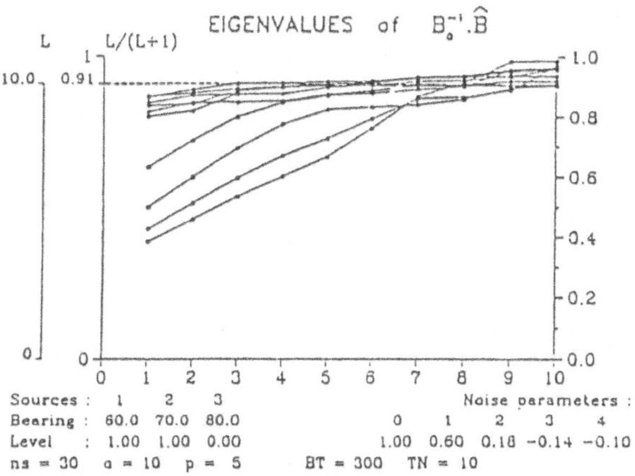

Fig. 4: 10 trials. Effect of the choice of L. L = 10

Obviously an efficient way in order to judge performances of the methods consists also in comparing the results of an HR method (MUSIC for instance) before and after whitening. The whitening being achieved by means of the estimated noise model, the corresponding procedure is described below:

1) factor $\hat{B}_L$ in Choleski factors ($\hat{B}_L$ estimated noise CSM): $\hat{B}_L = T_e \cdot T_e^*$

2) compute the whitened matrix: $\hat{R}_W = T_e^{-1} \cdot \hat{R} \cdot T_e^{-1*}$

3) compute the function f (θ) of bearing θ, defined below:

$$f(\theta) = \frac{\|T_e^{-1} \cdot D_\theta\|^2}{\|T_e^{-1} \cdot D_\theta\|^2 - \sum_{i=1}^{s'} |V_i^* \cdot T_e^{-1} \cdot D_\theta|^2} \tag{3.37}$$

where:

$\begin{cases} s' \text{ is the assumed source number} \\ \{V_i\}_{i=1}^{s'} \text{ are the eigenvalues of } \hat{R}_W \text{ associated to its s' greatest eigenvalues } \lambda_1 \geq \ldots \geq \lambda_{s'} \end{cases}$

As it can be seen on fig. 5, the spatial coloration of noise induces many degradations to the MUSIC method's results. Thus, the two sources at 35 deg. and 42 deg. bearings are not separated (the source at 42 deg. is never seen before whitening) and spurious peaks (related to noise coloration) can be seen near the array broadside (90 deg.). The fig. 6 presents the improvements due to whitening, the two sources at 35 and 42 deg. are well separated whereas spurious peaks are cancelled.

The simulation parameters are given below:

- 16 sensors (linear array, equispaced),
- 3 sources (linear levels): 0.3 (35 deg.), 0.1 (42 deg.), 0.5 (70 deg.),
- number of assumed sources s' = 5,
- Toeplitz noise CSM: $B = \Sigma\, \beta_i \left(U_i + U_i^t\right)$

$$\beta_1 = 2.5\ ,\ \beta_2 = 1.5\ ,\ \beta_3 = 0.9\ ,\ \beta_4 = 0.4$$

- 100 snapshots (BT = 100)
- 7 trials.

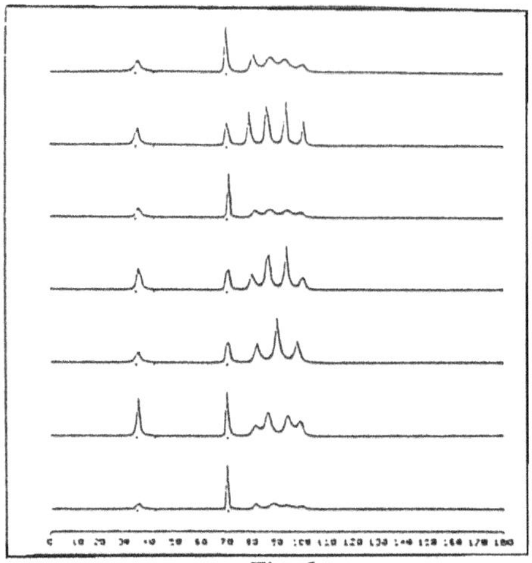

Fig. 5

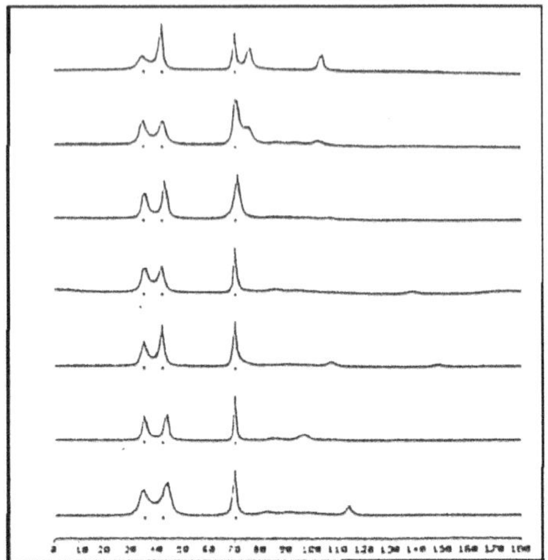

Fig. 6

Results of MUSIC method for 7 trials of the same simulation before whitening (fig. 5) and after (fig. 6).

3.3 - THE MAXIMUM LIKELIHOOD METHOD [13], [23]

3.3.1 - Introduction

Under the hypothesis of an additive white noise, the lowest eigenvalues of the observation matrix R are asymptotically equal. This is the basic idea of the proposed method which consists mainly in clustering the lowest eigenvalues of the "whitened" observation matrix $\widehat{R}_W$ w.r.t. a noise model. An AR(MA) model is the more relevant for that purpose and will lead to simple and efficient algorithms. Actually, the method relies upon three factors:

F_1 - the likelihood functional may be expressed as a functional of the eigenvalues of the whitened covariance matrix (of the outputs),

F_2 - the inverse B^{-1} of the noise covariance matrix admits an explicit formulation in terms of the AR(MA) coefficients of the noise model,

F_3 - the derivatives of the likelihood functional may be easily calculated using (F_1) and (F_2) and classical results for eigenvalue perturbations.

Consider, first, the white noise case; assume furthermore the source number be equal to q.

Let $\{X_1, \ldots, X_N\}$ be a sequence of N independent complex gaussian vectors (snapshots vectors) with covariance matrix R, then the conditional density of this sequence is:

$$f\left(X_1, \ldots, X_N / \lambda, S\right) = \pi^{-N \cdot n_s} \cdot (\det R)^{-N} \cdot \exp\left(-N \operatorname{tr}\left(\widehat{R} \cdot R^{-1}\right)\right)$$

where:

$$\begin{cases} \widehat{R} = \frac{1}{N} \cdot \sum_{i=1}^{N} X_i \cdot X_i^{*} \\ R = S + \lambda \cdot Id \\ S\text{: being the unknown covariance matrix of sources (rank } S = q) \end{cases} \qquad (3.38)$$

We are now seeking the value of λ and the matrix S which maximize the likelihood. Forgetting the special structure of S (plane wave hypothesis), it is specified by its eigensystem [23]. A remarkable (and classical) fact is that the log-likelihood may be simply expressed as a functional of the eigenvalues of $\widehat{R}$, more precisely:

$$\sup f\left(X_1, \ldots, X_N / \lambda, S\right) = (n_s - q) \operatorname{Log} ge(q) - (n_s - q) \operatorname{Log} ar(q) + cst$$

where:

$$\begin{cases} ge(q) = \prod_{i=q+1}^{n_s} \left(\hat{\lambda}_i\right)^{\frac{1}{n_s - q}} & : \text{geometric mean} \\ ar(q) = \frac{1}{n_s - q} \cdot \sum_{i=q+1}^{n_s} \hat{\lambda}_i & : \text{arithmetic mean} \\ \hat{\lambda}_1 \geq \ldots \geq \hat{\lambda}_{n_s} & : \text{eigenvalues of } \hat{R} \end{cases} \tag{3.39}$$

Consider, now the case of an unknown correlated noise, then the likelihood functional $L_q(B, S)$ is given by:

$$L_q(B, S) = -\,\text{Log}\det R - \text{tr}\,(R^{-1} . \hat{R})$$

($R = S + \lambda . B$; λ scalar)

Then by elementary calculations [23], the following expression (rather similar to the previous one) of $L_q(B, S)$ is obtained:

$$L_q(B, S) = (n_s - q) . \log ge_w(q) - (n_s - q) . \log ar_w(q) + cst \tag{3.40}$$

where:

$ar_w(q)$ and $ge_w(q)$ are the arithmetic and geometric means of the whitened matrix $\hat{R}_w$ ($\hat{R}_w = B^{-1} . \hat{R}$).

Now the problem takes the following form:

"Maximization of $L_q(B)$ (for a given q) relatively to the parameters defining B"

Stress that this maximization is considered only w.r.t. B since S have been estimated conditionally to B by means of eigensystem estimation. A simple and effective way to solve it, will be exposed in the next sections; in fact the numerical problem relies heavily on the parametrization presented below.

3.3.2 - Noise parametrization - Application

Consider, first, an AR noise modelling, then the inverse of its covariance matrix has an explicit formulation given by a formula attributed to Gohberg [24]:

$$B^{-1} = \frac{1}{\sigma^2} . \left(A_1 . A_1^t - A_3 . A_3^t\right)$$

with A_1, A_3 triangular matrices given by:

$$\begin{cases} A_1 = \sum_{i=0}^{p} a_i . Z^i & \text{(by convention } a_0 = 1) \\ A_3 = \sum_{i=1}^{p} a_i . Z^{n_s - i} \end{cases} \tag{3.41}$$

In the above expression Z^i is the shifting matrix given by:

$$Z^i(j, k) = \begin{cases} 1 & \text{if } j - k = i \quad 1 \leq j, k \leq n_s \\ 0 & \text{else} \end{cases} \tag{3.42}$$

(note that: $Z^{i+1} = \left(Z^i\right) . Z$, $Z^0 = Id$)

The matrix B^{-1} may be written as a quadratic form w.r.t. the coefficients $\{a_i\}$, the matrix A_3 can be considered as a corrective term. Note, furthermore, that B^{-1} as defined above is centrosymmetric whereas its inverse (i.e.: B) is a Toeplitz matrix [24].

On the opposite, there is no explicit expression of B^{-1} for an MA modelling, however a simple expression of B is given by:

$$B = \left(\sum_{i=0}^{p} b_i . Y_i\right) . \left(\sum_{i=0}^{p} b_i . Y_i\right)^t$$

where $\{b_i\}_{i=0}^{p}$ are the MA coefficients and Y_i are rectangular (n_s x 2 n_s) matrices defined by:

$$Y_i\,(j,k) = \begin{cases} 1 & \text{if } k - j = i \\ 0 & \text{else} \end{cases} \tag{3.43}$$

A parametrization of B for ARMA modelling is obtained by combination of (3.41) and (3.43). The above parametrization will be convenient for maximizing the likelihood functional as we shall see now.

3.3.3 - Maximization of the likelihood functional

In general case, iterative methods are needed for maximization of L_q (B).

A general form of a gradient algorithm can be written as follows:

$$\begin{cases} A_{k+1} = A_k - \rho_k \,.\, G_k \\ A_k^t = \left(a_1^k, \ldots, a_p^k\right) \quad \text{the estimated parameters at the k-th iteration} \end{cases}$$

The problem consists now in calculating the gradient vector $G_k \left(G_k\,(i) = \frac{\partial}{\partial a_i} L_q\,(B) \right)$ For that purpose, the partial derivatives of the eigenvalues of $B_k^{-1}\,.\,\widehat{R}\ \left(\equiv \widehat{R}_{w,k}\right)$ are needed.

Assuming that these eigenvalues $\left(\lambda_j^k\right)$ are simple, they are given by the following formula [25]:

$$\frac{\partial}{\partial a_i}\lambda_j^k = \left(U_j^k\right)^* . \left(\Delta_i'^k\right) . \left(U_j^k\right)$$

where:

$$\begin{cases} .\ U_j^{\,k} \text{ being eigenvector associated to the eigenvalue } \lambda_j^k \text{ of } B_k^{-1}\,.\,\widehat{R} \\ \text{or } T^*\,.\,B_k^{-1}\,.\,T \quad \left(\widehat{R} = T\,.\,T^*\right) \\ .\ \Delta_i'^{\,k} = \frac{\partial}{\partial a_i}\left(T^*\,.\,B_k^{-1}\,.\,T\right) = T^*\,.\,\Delta_i^k\,.\,T \end{cases} \tag{3.44}$$

The Gohberg's formula allows us to obtain a very simple expression of the matrix Δ_i^k. The gradient vector G_k is straightforwardly deduced from the partial derivatives of the lowest eigenvalues.

Practically, a suitable estimation of the optimal step size is very useful in order to ensure convergence of the algorithm. A first order approximation is sufficient for most of the cases and is obtained as follows:

1) Define:

$$\lambda_j^k\,(\rho) = \lambda_j - \rho\,.\,G_k^t\,.\,\Lambda_j^k$$

with

$$\left(\Lambda_j^k\right)^t = \left(\frac{\partial}{\partial a_1}\lambda_j^k, \ldots, \frac{\partial}{\partial a_p}\lambda_j^k\right) \tag{3.45}$$

2) Substitute λ_j^k (ρ) defined by (3.45), (3.44), (3.41) in the likelihood functional L_q and by means of an unidimensional method (of maximization) determine ρ.

The parametrization of the noise matrix B may be advantageously expressed in terms of reflection coefficients [23], using this way the stability of the estimated AR model (of noise) is easily ensured. However, if the noise poles approach the unit circle then the noise covariance matrix tends towards singularity and, therefore, the functional approaches - ∞. When the step size is not too important this fact avoids stability problems.

3.3.4 - Convergence analysis

Consider, for instance, a given AR model (order 2), it is then possible to compute the functional $L_{a_1, a_2}(q)$. Then the maximum of the functional is reached for the exact values of AR parameters but it is not concave on the whole domain [13], [23].

The likelihood functional (defined by (3.40)) enjoys however the following properties (regardless estimation problems of $\widehat{R}$ and assuming an appropriate noise model).

Property 1: *The likelihood functional defined by (3.40) is negative, it is null if the noise model is perfectly estimated.*

This is a simple consequence of the properties of arithmetic and geometric means [24], the nullity of the functional is equivalent to the equality of the $(n_s - q)$ lowest eigenvalues of $\widehat{R}_w$.

Property 2: *The gradient vector of the log-likelihood functional is null if the vector of AR parameters A is equal to A_0 (exact values of the parameters).*

A direct proof of property 2 based on the analysis of the eigenvalues seems very difficult or, at best, very cumbersome. Therefore, a direct approach is preferred.

Using classical results of linear algebra, one obtains:
$$\frac{\partial}{\partial a_i} \log \det R = \mathrm{tr}\left(R^{-1} . \frac{\partial}{\partial a_i} . B\right)$$
$$= - \mathrm{tr}\left(R^{-1} . B . \Delta_i . B\right)$$

so that, finally:
$$\frac{\partial}{\partial a_i} L(q, A) = \mathrm{tr}\left[B . R^{-1} . \left(R - \widehat{R}\right) . R^{-1} . B . \Delta_i\right] \tag{3.46}$$

Therefore, nullity of the gradient vector implies:
$$\sum_{i=1}^{p} a_i . \frac{\partial}{\partial a_i} L(q, A) = \mathrm{tr}\left[\left(Id - \widehat{R} . R^{-1}\right) . B . R^{-1}\right]$$
$$= \mathrm{tr}\left[T^* . R^{-1} . \left(R - \widehat{R}\right) . R^{-1} . T\right] = 0 \tag{3.47}$$

(with $B = T . T^*$)

Now, it is always possible to choose starting value of AR parameters and step sizes in order to assume the positivity of the matrix $\left(\widehat{R} - R\right)$ which amounts (by definition of the likelihood functional) to positivity of the matrix $B_0 - B$ $(R \equiv S + B_0)$.

Consequently, nullity of the gradient vector implies equality of exact and estimated noise parameters.

Examine now the convergence of the gradient method, the following inequality is easily verified:

$$\left\| \nabla L_q(A) - \nabla L_q(A') \right\| \leq C \, \| A - A' \|$$

(C : positive constant, ∇ denoting the gradient vector)

Consider now the functional $L'(A) = - L_q(A)$, then:

$$L'(A) - L'(A') = \int_0^1 \left\langle \nabla L'(A' + \tau (A' - A), A - A' \right\rangle d\tau$$

$$= \left\langle \nabla L'(A), A - A' \right\rangle + \int_0^1 \left\langle \nabla L'(A + \tau (A - A') - \nabla L'(A), A - A' \right\rangle d\tau$$

$$\geq \langle \nabla L'(A), A - A' \rangle - \int_0^1 \| \nabla L'(A + \tau(A - A') - \nabla L'(A) \| . \|A - A'\| \, d\tau$$

using Cauchy-Schwarz inequality ($\langle \nabla L', A \rangle$ denoting the classical scalar product $\sum a_i . \frac{\partial}{\partial a_i} L'$).

Therefore, the following inequality is obtained:

$$L'(A) - L'(A') \geq \langle \nabla L'(A), A - A' \rangle - C'. \|A - A'\|^2$$

Consider now the gradient algorithm, then one obtains finally:

$$L'(A_k) - L'(A_{k+1}) \geq (1 - C'. \rho_k). \rho_k. \|\nabla L'(A_k)\|^2 \tag{3.48}$$

Consequently, the sequence $\{L'(A_k)\}$ decreases (as k increases) if the step size is sufficiently small (i.e. : inferior to 1/C'). Now, by property 1, this sequence is strictly positive hence it converges so that finally, using property 2, the vector A_k converges towards A_0.

Obviously, the errors on the estimation of $\widehat{R}$ leads to statistical results but the gradient method still converges and gives very satisfying results as it will be shown later.

Some parameters have been fixed for the previous derivations, these are q (number of sources) and p (AR order). In practice they are unknown, we shall examine now the consequences of misadjustments of these parameters.

A) Number of sources

Property 1 proves that it is (theoretically) possible to **overdetermine** the source number so far as p (AR noise model) remains inferior to (n_s - q), the estimation errors on $\widehat{R}$ lead to slightly degrade the results (random eigenvalues) but the degradation is quite acceptable.

The "classical" information criteria (Akaike, MDL, ... [26]) do not provide a satisfying source number estimate, because they use eigenvalues of $\widehat{R}$ and do not separate sources and correlated noise; for instance, the estimated source number (by means of these criteria) is maximal (i.e.: sensor number) for most of the simulations presented in [13], [23]. However it is possible to remedy this problem by some means, among them are :

1) Modifications of information criteria [23]

   They consist mainly in adding the noise model parameters in the set of "free" parameters [23]. The statistical results of the corresponding criterion are satisfying however its computation cost can be rather important.

2) Statistical properties of the eigenvalues of a "corner" of the matrix $\widehat{R}$ (MA case)

   This original approach has been recently presented by J.J. Fuchs [27] and gives an accurate estimate of the source number in presence of correlated noise.

3) Use of a state space approach for sensor output modelling (ARMA case)

B) Noise model order

In the asymptotic case, noise model overdetermination does not affect the result (e.g.: in terms of spatial noise density) as it was shown by the proof of property 2.

For all simulated data, it seems possible to overdetermine the noise model without dramatic effects. The robustness of the method to misadjustments of these parameters is indeed a crucial point [13], [23].

Obviously, the parameters p and q may be estimated simultaneously as shown previously (A.1), but a separate estimation of the order p seems rather difficult.

3.3.5 - Simulation results

The first result (fig. 7) corresponds to an AR modelling, the simulation data are given in the caption of fig. 7. Note that the dimension of source subspace is 5 (i.e.: $q = 5$) whereas the results are quite satisfying. The broken line corresponds to MUSIC results without whitening ($\widehat{R}$ is Toeplitz) while the solid line corresponds to results after whitening (corresponding to the noise estimated model). The two sources at bearings 35 and 45 deg. are not separated whereas the weak source is not detected, without whitening. Conversely, the whitening yields the resolution of the two strong sources and allows to "detect" the weak one.

Fig. 8 represents the noise spatial density associated to the model. Obviously, its peak (at 90 deg.) yields the spurious peaks (at the vicinity of 90 deg.) and non-detection of the weak source (110 deg.) for the unwhitened data. The broken line (exact noise density) and the solid line estimated noise density) are perfectly merged.

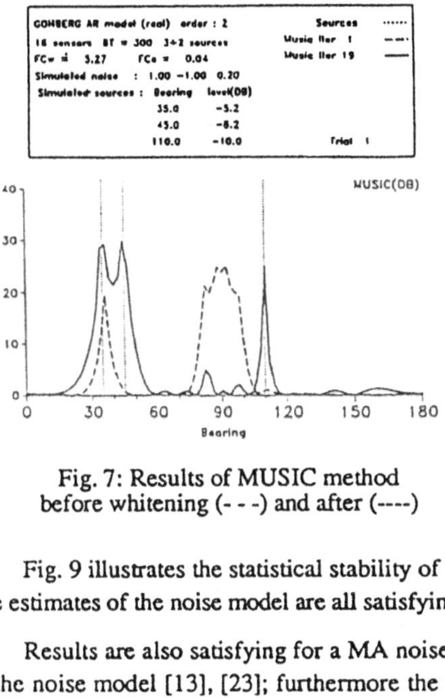

Fig. 7: Results of MUSIC method before whitening (- - -) and after (----)

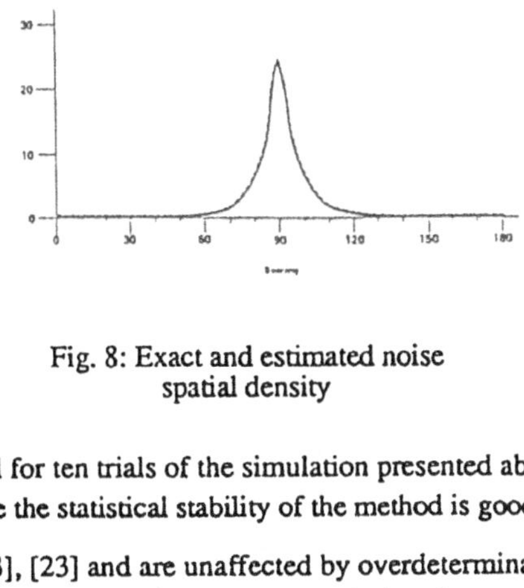

Fig. 8: Exact and estimated noise spatial density

Fig. 9 illustrates the statistical stability of the method for ten trials of the simulation presented above. The estimates of the noise model are all satisfying, therefore the statistical stability of the method is good.

Results are also satisfying for a MA noise model [13], [23] and are unaffected by overdetermination of the noise model [13], [23]; furthermore the method can be straightforwardly extended to the complex case as shown at fig. 10.

For all simulation results, the algorithm (maximization of the likelihood functional) starts with $A^t = (1, 0, \ldots, 0)$ and converges fast (at most ten iterations). This algorithm has also been tested on real data (at sea) with satisfying results [13].

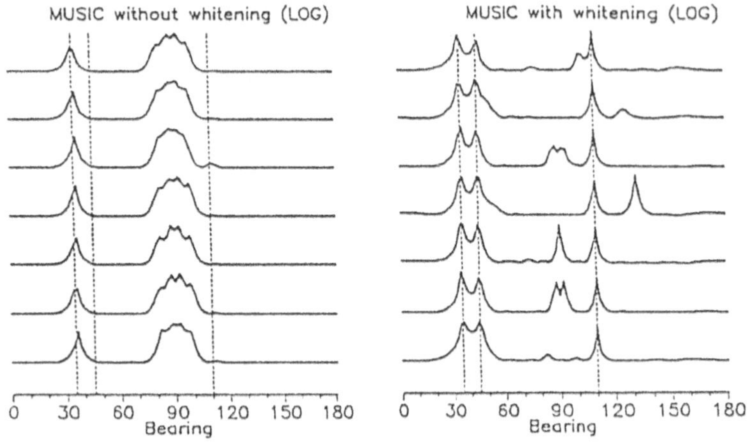

Fig. 9: Simulation parameters identical to figures 7 and 8.
Statistical behavior of the method on 7 trials.

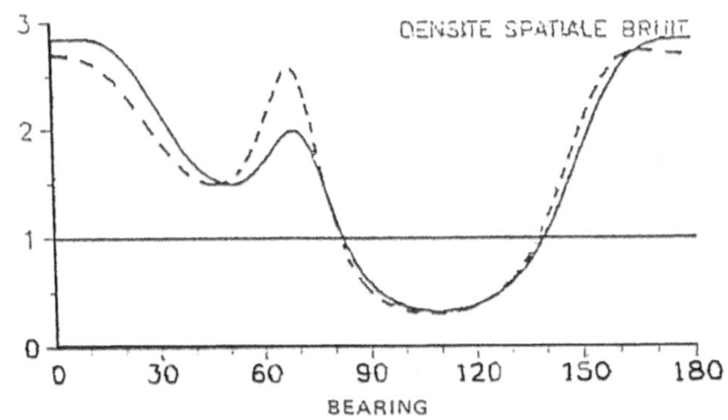

Fig. 10: Complex case. Comparison of exact noise spatial density (----) and its estimates.

3.4 - PLANE WAVE HYPOTHESIS

Conversely to the preceding approaches, special structures of source vectors (plane wave hypothesis) can be considered. Taking into account plane wave hypothesis leads to define a functional depending upon both source and noise parameters.

Consider the case of an AR noise modelling, let A be the vector of coefficients of this model (i.e. $A^t = (1, a_1, \dots, a_p)$) and $\left(A(z^{-1})\right)^{-1}$ the associated filter. The whitening filter is thus $A(z^{-1})$ and the covariances of whitened data (using this filter) can be written as:

$$r_w(l) = \mathbb{E}\left\{\left(\sum_{i=0}^{p} a_i\, x_{t-i}\right) \cdot \left(\sum_{i=0}^{p} \bar{a}_i \cdot \bar{x}_{t-l-i}\right)\right\} \qquad a_0 = 1$$

$$= A^{*} . R_l . A$$

where R_l is the $(p + 1 \times p + 1)$ matrix (extracted from R) given by:

$$R_l = \begin{pmatrix} r(l) & r(l-1) & \cdots & r(l-p) \\ \vdots & \ddots & \ddots & \vdots \\ \vdots & & \ddots & r(l-1) \\ r(l+p) & \cdots & \cdots & r(l) \end{pmatrix} \quad (3.49)$$

Define now the functional (conditionally to A):

$$J_w(A) = \sum_{l=1}^{L} |r_w(l) - r_s(l)| \quad (3.50)$$

where covariances $r_w(l)$ are defined by (3.49) and $r_s(l)$ are associated to sources.

Now, for our applications, source parameters are unknown; leading to define a functional called "whiteness functional" by:

$$J_w(A) = \sum_{l=1}^{L} |r_w(l) - \hat{r}_s(l)|^2 \quad (3.51)$$

where $\hat{r}_s(l)$ is estimated by means of Pisarenko's method (or any HR method) applied to whitened data (conditionally to A), i.e.:

$$\hat{r}_s(l) = \sum_{i=1}^{s} \hat{q}_i \,.\, \exp\left(-j\pi l \hat{f}_i\right) \qquad j^2 = -1 \quad (3.52)$$

($\hat{q}_i$, f_i : power and spatial frequency to the i-th source)

Then, the method consists in adjusting the parameter vector A in order to minimize the functional J_w (A), this is achieved by means of a gradient method. The main difficulty consists in calculation of partial derivatives of $\hat{f}_i$ w.r.t a_j by means of perturbation analysis of the projection function, this is detailed in [23].

In fact, the method relies upon the whitening invariance (of bearings) properties. More precisely, formula (3.49) defining $r_w(l)$ may be written in matricial form, i.e.: $R_w = A R A^*$

where R ($n_s \times n_s$) is the Toeplitz covariance matrix of the outputs, and A a rectangular ($L \times n_s$) matrix ($L = n_s - p$) defined by:

$$A = \begin{pmatrix} 1 & a_1 & \cdots & a_p & 0 & \cdots & \cdots & 0 \\ 0 & \ddots & & & \ddots & & & \vdots \\ \vdots & \ddots & \ddots & & & \ddots & & \vdots \\ 0 & \cdots & 0 & 1 & a_1 & \cdots & a_p & 0 \cdots 0 \end{pmatrix} \quad (3.53)$$

We shall consider now the effect of whitening (formula (3.53) on a plane wave whose associated steering vector is denoted D_θ: $D_\theta^t = (1, \exp(j\alpha), \ldots, \exp(j(n_s - 1)\alpha)$ $\qquad j^2 = -1$

then using (3.53) the covariance matrix of the source after whitening is given by:

$$R_w(\theta) = A D_\theta D_\theta^* A^*$$

now:

$$A D_\theta = \left(1 + a_1 e^{j\alpha} + \ldots + a_p e^{j(p-1)\alpha}\right) D_\theta'$$

D_θ' : being the ($L \times 1$) vector obtained by selecting the L first components of the vector D_θ.

therefore:

$$R_w(\theta) = q(\theta)\, D_\theta'\; D_\theta'^*$$

with: $\quad q(\theta) = \left|1 + a_1 e^{j\alpha} + \ldots + a_p e^{j(p-1)\alpha}\right|^2 \quad (3.54)$

The whitening procedure defined by (3.53) enjoys the following properties:

1) Let R be a Toeplitz covariance matrix, then $R_w = A R A^*$ is also a Toeplitz matrix.

2) The procedure transforms a steering vector in another steering vector associated to the same bearing.

These properties are, in fact, fundamental because they justify the whiteness functional definition and the calculation of the gradient. In practical situations, exact source bearings are unknown, the above properties stand only that these bearings are unaffected by the whitening procedure.

This method gives satisfying results [13, 23], but is less robust than the two previous methods.

4 - NOISE WHITENING PROCEDURES

Along the paper, some procedures for noise whitening have been considered. Let us detail them now.

The first method (3.37) consists in whitening by calculation of triangular factors of the noise CSM (i.e. $B = T . T^*$ and $R_w = T^{-1} . R . T^{-1\,*}$). Its main advantage is simplicity and the results obtained by this way are satisfying; however the Toeplitz structure of R is not preserved.

In order to remedy this problem it is possible to use the inverse filtering approach presented in § 3.4. It relies mainly upon the simplicity of the inverse filter in the case of an AR noise modelling. It preserves Toeplitz structure and furthermore it avoids computation of triangular factors ($B = T . T^*$) which may be very interesting for large arrays.

5 - STATE SPACE REPRESENTATION OF THE SENSOR OUTPUTS

In the previous sections, it is shown that "usual" HR methods are not robust w.r.t. the noise correlation, and some methods for noise model estimation are presented in order to remedy this problem.

For an equally spaced line array, there is another way.

Indeed, the estimation of the source subspace is the cause of the problem, and we intend here to take care of it thanks to a state space representation of the sensor outputs, which is possible with an array of this type.

The first considered approach deals with approximated stochastic realization methods. The solution which is derived is satisfactory because the robustness w.r.t. the unknown noise correlation is considerably enhanced relatively to usual HR methods; but, this solution is obtained via principal component analysis, which does not allow to take into account the special structure of the source subspace induced by plane wave and spatial stationarity hypotheses.

The theory of approximation of rational functions in Hardy spaces, developed by Adamjan, Arov and Krein, is the foundation of the solution of this fundamental problem.

This theoretical problem posed in infinite dimension is translated into a finite dimension one by using some properties of linear systems (balanced realization), and leads to a solution for array processing which is still quite robust w.r.t. the unknown noise correlation, but also which improves the detection of weak sources in white noise w.r.t. usual methods, owing to the preservation of the problem structure (and the use of L_∞ norm).

Both approaches rely upon a state space modelling of the sensor outputs which is necessary to introduce now.

5.1 - INTRODUCTION OF THE STATE SPACE REPRESENTATION

The considered problem consists in determining the bearings $\{\theta_k\}_{k=1,\dots,m}$ of m waves impinging a line array composed with n equally spaced sensors.

Under the **plane wave hypothesis**, the difference in phase between two contiguous sensors relative to a wave is constant and connected to the source bearing by:

$$\varphi_k = 2\pi f \frac{d}{c} \cos\theta_k \tag{5.1}$$

for the k-th source, where f is the frequency of interest, d is the distance between two contiguous sensors and c is the velocity of wave.

So the problem amounts to estimate the phases φ_k which represent the deterministic propagation of the plane waves between two sensors, from the sensor outputs.

At a given frequency f (omitted for more convenience), the output of the i-th sensor is given by:

$$y_i = \sum_{k=1}^{m} \alpha_k \exp[-j(i-1)\varphi_k] + v_i \qquad j^2 = -1 \tag{5.2}$$

where α_k is the random level of the k-th source, and $\{v_i\}$ is a gaussian mean zero additive noise.

It is now interesting to show the ability of a state space model to describe the sensor outputs and to suit to the noise model.

5.1.1 - Additive spatially white noise case

It is thought that the noise process $\{v_i\}$ is spatially white, gaussian, mean zero.

So, the process $\{y_i\}$ can be represented as the sum of a "spatial" autoregressive (AR) process of order m which poles are on the unit circle, and a spatially white gaussian mean zero noise.

Then, when the propagation model is perfect, the set of poles is $\{\exp[-j\varphi_k],\ k=1,\dots,m\}$, which means that the propagation simply results in phasing factors (plane wave hypothesis). Due to small imperfections, the poles are close to the unit circle but inside it (stability), and the set of poles is $\{\rho_k \exp[-j\hat{\varphi}_k],\ k=1,\dots,m\}$ where ρ_k is close to unity and $\hat{\varphi}_k$ is close to the phasing factor relative to the k-th source.

A rank m linear state space model of the y-process is given by the pair of equations:

$$\begin{cases} X_{i+1} = F\,X_i \\ y_i = \mathbf{h}^* X_i + v_i \end{cases} \tag{5.3}$$

where y_i is the output of the i-th sensor, X_i is the state vector, v_i is the additive noise, F and $\mathbf{h}$ are deterministic parameters.

The second equation is the observation equation, whereas the first one is called the state transition equation and provides a prediction of the state vector at the next sensor. So, the matrix F has to describe the evolution parameters between two contiguous sensors. The size of this matrix equals the dimension of the state, and the minimal dimension of the state is equal to the number of parameters which are necessary to describe the process evolutions between two sensors.

Here, the predictable part of the received signal is the sum of m plane waves; so that, the minimal dimension of the state is equal to m, and the m x m transition matrix F contains the phasing factors relative to the m plane waves. The poles of the y-process are the eigenvalues of the matrix F, that means that it exists a basis in which $F = \text{diag}(\exp[-j\varphi_k])$.

Therefore, the determination of the source bearings leads to determine the transition matrix of a given rank linear system. This problem is well-known in the field of linear systems and is called "realization problem". Its resolution requires the estimation of the covariances of the process $\{y_i\}$.

The sensor outputs being modelled by the linear system (5.3), their covariances $r_p = E\left(y_{i+p}\, y_i^*\right)$, under stationarity hypothesis, are given by:

$$r_p = h^* F^p\, P\, h + b^2\, \delta(p) \qquad \text{if } p \geq 0 \tag{5.4}$$

and

$$r_{-p} = r_p^*$$

with $P = E\left(X_i\, X_i^*\right)$, $b^2 = E\left(v_i\, v_i^*\right)$ and $\delta(p) = \begin{cases} 1 & , \text{if } p = 0 \\ 0 & , \text{if } p \neq 0 \end{cases}$ (P is the state variance matrix).

The estimation of the transition matrix F from these covariances will be presented later.

5.1.2 - Additive spatially correlated noise case

The process $\{y_i\}$ can still be represented as the output of the above state space model (5.3) where the observation noise $\{v_i\}$ is spatially correlated now.

Unfortunately, this model is useless to extract the interesting parameters because there is no efficient realization methods which are suited to this model with a correlated noise on the observation equation. Nevertheless, another state space representation of the process $\{y_i\}$ can be considered in the additive correlated noise case.

Indeed, it is known that the observation noise can be well approximated by a spatial autoregressive moving average (ARMA) process, whereas the sum of plane waves is still modelled by an AR process.Therefore, the sum of these processes can be represented by an ARMA process of order greater than the plane wave number. In terms of linear systems, the received signal y_i, which is the sum of a source part s_i with state space representation

$$\begin{cases} X_{i+1}^s = F_s\, X_i^s \\ s_i = h_s^*\, X_i^s \end{cases}$$

and a noise part v_i with state space representation

$$\begin{cases} X_{i+1}^v = F_v\, X_i^v + T_v\, w_i \\ v_i = h_v^*\, X_i^v + w_i \end{cases}$$

can be modelled by a state space model as follows:

$$\begin{cases} \begin{pmatrix} X_{i+1}^s \\ X_{i+1}^v \end{pmatrix} = \begin{pmatrix} F_s & 0 \\ 0 & F_v \end{pmatrix} \begin{pmatrix} X_i^s \\ X_i^v \end{pmatrix} + \begin{pmatrix} 0 \\ T_v \end{pmatrix} w_i \\ y_i = \begin{pmatrix} h_s^* & h_v^* \end{pmatrix} \begin{pmatrix} X_i^s \\ X_i^v \end{pmatrix} + w_i \end{cases}$$

This can be written as:

$$\begin{cases} X_{i+1} = F\,X_i + T\,w_i \\ y_i = h^*\,X_i + w_i \end{cases} \tag{5.5}$$

which is the innovation model of an ARMA process.

Here, the minimal dimension of the state has to be increased relatively to the source number to take into account the evolution parameters of the noise.

Therefore, the pole number of the y-process is greater than the plane wave number. Among them, m poles are close to the unit circle and correspond to the phasing factors of the plane waves $(\exp[-j\,\varphi_k],\ k = 1, \ldots, m)$, the others lie more within the unit circle (stability) and correspond to the noise. All these poles are still the eigenvalues of the transition matrix F, and the problem reverts to a realization problem which requires the estimation of the observed process covariances. In terms of the state space parameters the sensor output covariances are given by:

$$r_p = \begin{cases} h^*\,P\,h + b^2 & , \text{ if } p = 0 \\ h^*\,F^{p-1}\left(F\,P\,h + b^2\,T\right) & , \text{ if } p > 0 \end{cases}$$

and

$$r_{-p} = r_p^* \tag{5.6}$$

with $P = E\left(X_i\,X_i^*\right)$ and $b^2 = E\left(w_i\,w_i^*\right)$ (P is the state variance matrix).

From this model, several methods can be applied to obtain the interesting poles of the observed process.

Remark: another interpretation to introduce the innovation model of the observed process is possible in terms of spatial densities.

Indeed, under the assumption that the sources and the noise are statistically independent, the spatial density of the process $\{y_i\}$ is the sum of the spatial density of the source part and the spatial density of the noise part, i.e. the sum of an AR spectrum and an ARMA one:

$$P_y = P_s + P_v = \frac{1}{A_s(z)\,A_s^*(1/z)} + \frac{B_v(z)\,B_v^*(1/z)}{A_v(z)\,A_v^*(1/z)}$$

Then, P_y is given by:

$$P_y = \frac{A_v(z)\,A_v^*(1/z) + A_s(z)\,A_s^*(1/z)\,B_v(z)\,B_v^*(1/z)}{A_s(z)\,A_s^*(1/z)\,A_v(z)\,A_v^*(1/z)}$$

So, the process $\{y_i\}$ has a non-negative rational spatial density which means it can be represented by an innovation model (5.5).

5.2 - REALIZATION PROBLEM

The estimation of the source bearings by means of state space models leads to consider the realization problem which consists in determining the parameters of a fixed rank linear system, especially the transition matrix, whose output covariances best approximate the sensor output covariances [27], [28].

Two approaches of this problem are investigated here: the approximated realization method and the optimal realization method.

Before dealing with the case of an unknown additive correlated noise, the simple case of an additive white noise is considered.

5.3 - SPATIAL REALIZATION METHOD, ADDITIVE WHITE NOISE CASE

A rank m linear state space model of the sensor outputs is given in the case of an additive white noise by (5.3), and its covariances are given by (5.4).

Let Y be the output vector of the array: $Y = (y_1 \; y_2 \dots y_n)^T$

and R = E (Y Y*) the exact covariance matrix of the sensor outputs:

$$R = \begin{pmatrix} r_0 & r_{-1} & \cdots & r_{-n+1} \\ r_1 & \ddots & \ddots & \vdots \\ \vdots & \ddots & \ddots & r_{-1} \\ r_{n-1} & \cdots & r_1 & r_0 \end{pmatrix}$$

Using (5.4), the output covariance matrix can be written as: $R = \mathcal{O} P \mathcal{O}^* + b^2 I_n$ (5.7)

where $\mathcal{O}$ is the n x n observability matrix of the system [29], [30]:

$$\mathcal{O} = \begin{bmatrix} h^* \\ h^* F \\ \vdots \\ h^* F^{n-1} \end{bmatrix}$$

The following property can be easily deduced from the structure of the observability matrix:

$$\mathcal{O}^{\uparrow} F = \mathcal{O}^{\downarrow} \tag{5.8}$$

where $\mathcal{O}^{\uparrow}$ and $\mathcal{O}^{\downarrow}$ are (n - 1) x m matrices, ↑ selects the (n - 1) first rows of a matrix, and ↓ selects the (n - 1) last rows of a matrix.

Therefore, the transition matrix can be derived as: $F = \left(\mathcal{O}^{\uparrow}\right)^{\#} \mathcal{O}^{\downarrow}$ (5.9)
where # stands for pseudo-inverse.

This method was initially introduced by S.Y. Kung and al. [31] as the Toeplitz Approximation Method (TAM), and then connected to the array processing field [32] as follows:

Let $\mathcal{D}$ be the n x m source steering matrix:

$$\mathcal{D} = \begin{pmatrix} 1 & \cdots & 1 \\ \exp[-j\,\varphi_1] & \cdots & \exp[-j\,\varphi_m] \\ \vdots & \vdots & \vdots \\ \exp[-j\,(n-1)\,\varphi_1] & \cdots & \exp[-j\,(n-1)\,\varphi_m] \end{pmatrix}$$

then the covariance matrix can also be written in terms of array processing as:

$$R = \mathcal{D} \gamma \mathcal{D}^* + b^2 I_n \tag{5.10}$$

where γ is the m x m source covariance matrix.

In the eigenbasis of F, the plane wave hypothesis yields: $\mathcal{D}^{\uparrow} F = \mathcal{D}^{\downarrow}$ (5.11)

so,

$$F = \left(\mathcal{D}^{\uparrow}\right)^{\#} \mathcal{D}^{\downarrow} \tag{5.12}$$

Therefore, an estimation of $\mathcal{D}$ is required.

Let the eigendecomposition of the covariance matrix be given by:

$$R = [U_1 \ U_2] \begin{bmatrix} \Sigma_1^2 & 0 \\ 0 & \Sigma_2^2 \end{bmatrix} \begin{bmatrix} U_1^* \\ U_2^* \end{bmatrix} \tag{5.13}$$

where the eigenvalues are in decreasing order and the subscript 1 stands for the dominant components. Then, its best rank m approximate is $U_1 \ \Sigma_1^2 \ U_1^*$.

Under white noise assumption, the power spectral density of the noise is equal to the arithmetical mean of the (n - m) smallest eigenvalues. By subtracting this estimate to the dominant eigenvalues, the best noise free rank m estimation of the covariance matrix is derived: $U_1 \ \hat{\Sigma}_1^2 \ U_1^*$

Comparing with the signal part of (5.10), one obtains: $\mathscr{D} \gamma \mathscr{D}^* = U_1 \ \hat{\Sigma}_1^2 \ U_1^*$ (5.14)

The use of the SVD for both matrices ($\mathscr{D} \gamma^{1/2}$) and ($U_1 \ \hat{\Sigma}_1$) provides: $\mathscr{D} \gamma^{1/2} = U_1 \ \hat{\Sigma}_1 \ W^*$ (5.15)
where W is a m x m unitary matrix:

Using the equations (5.14), (5.15) and the following property:

Property: *let M_1, M_2 be l x m and m x n matrices then:*

$$(M_1 \ M_2)^{\uparrow} = M_1^{\uparrow} \ M_2 \quad \text{and} \quad (M_1 \ M_2)^{\downarrow} = M_1^{\downarrow} \ M_2$$

the transition matrix is derived:

$$F = \gamma^{1/2} \ W \ \hat{\Sigma}_1^{-1} \left[\left(U_1^{\uparrow}\right)^{\#} \ U_1^{\downarrow} \right] \hat{\Sigma}_1 \ W^* \ \gamma^{-1/2} \tag{5.16}$$

This matrix can not be calculated because the matrices γ and W are unknown; nevertheless, only the eigenvalues of the matrix F (i.e. the poles of the process) are interesting here, and by means of matrix determinant properties, it can be easily shown that these eigenvalues are identical to these of the matrix $\left[\left(U_1^{\uparrow}\right)^{\#} \ U_1^{\downarrow} \right]$.

Therefore, the knowledge of the dominant eigenvectors of R is sufficient to estimate the bearings. In practical situation, the matrix R is replaced by its estimation , which provides an estimate of U_1 and then an estimate of the poles.

However the source number is generally overestimated and a separation between true poles (corresponding to real sources) and spurious ones is required. This can be achieved by taking into consideration the modulus of the pole estimates. Indeed, a modulus close to unity corresponds to a real source, whereas the poles of the noise lie more within the unit circle.

This method enhances the detection of weak sources in additive white noise, but is still sensitive to the noise correlation. A more important improvement is possible by taking into consideration the model of the noise.

Remark : the stability problem has been considered [9], and it has been shown that the equation (5.11) (or (5.8)) can be solved with a marginal stability constraint. Elsewhere, a simple and more efficient way to obtain a stable solution is achieved by orthogonally projecting the covariance matrix onto the space of Toeplitz matrices, which cancels at the same time the possible source cross-correlation (spatial smoothing).

5.4 - APPROXIMATED STOCHASTIC REALIZATION METHOD, ADDITIVE CORRELATED NOISE CASE

It is intended to take into account the spatial cross-correlation of the noise by modelling the sensor outputs as the outputs of a minimum phase innovation model (5.5). Then the approximated realization methods rely upon considerations of two vectors extracted from the array output vector:

$$Y_- = \left(y_{n_2}\ y_{n_2-1} \cdots y_1\right)^T$$
$$Y_+ = \left(y_{n_2+1}\ y_{n_2+2} \cdots y_n\right)^T \tag{5.17}$$

$(n_2 = \frac{n}{2}$, n even)

Y_- and Y_+ are respectively the past and the future of the observations at the n_2-th sensor. Let R_+, R_-, H be the covariances and Hankel matrices defined as follows:

$$R_+ = E\left(Y_+\ Y_+^*\right),\ R_- = E\left(Y_-\ Y_-^*\right) \text{ and } H = E\left(Y_+\ Y_-^*\right)$$

The aim of these methods consists in determining a state vector X of a minimum phase innovation model which sums up the past information Y_- in order to best predict the space of the future observations Y_+.

Thanks to the minimum phase hypothesis, the inverse filter of the state space model (5.5) is stable, and its state transition equation indicates that:

$$X = A\,Y_- \tag{5.18}$$

with $A = \left[T,\ \left(F - T\,h^*\right)T,\ \ldots,\ \left(F - T\,h^*\right)^{n_2-1}\ T\right]$

Therefore, the determination of the state vector X requires to find the matrix A which satisfies a given criterion. It is important to note that this matrix has theoretically a special structure. To solve this problem, two criteria are proposed.

5.4.1 - The predictive efficiency criterion [33], [34]

This criterion consists in determining the p x 1 state vector X ($p \geq m$) as a linear combination of the past observations, $X = A\,Y_-$, and which minimizes the prediction error variance of the future observations:

$$\underset{X}{\text{Min}}\left\{\text{tr}\left[\text{cov}\left(Y_+ - Y \mid X\right)\right]\right\} \tag{5.19}$$

where $\alpha \mid \beta$ stands for the orthogonal projection of the random variable α onto the space spanned by the random variable β, and is defined as follows:

$$\alpha \mid \beta = E\left(\alpha\,\beta^*\right)\left[E\left(\beta\,\beta^*\right)\right]^{-1}\beta$$

Using the relation (5.18) and the orthogonal projection theorem, the problem consists then in determining the p x n_2 matrix A which minimizes the following criterion:

$$\underset{A}{\text{Min}}\left\{\text{tr}\left[R_+ - H\,A^*\left(A\,R_-\,A^*\right)^{-1} A\,H^*\right]\right\} \tag{5.20}$$

The solution of this problem requires the eigendecomposition of the matrix $\left(H\,R_-^{-1}\,H^*\right)$ and is given by [9], [11]:

$$A = B\ \Sigma_1^{-1}\ U_1^*\ H\,R_-^{-1} \tag{5.21}$$

where B is any p x p invertible matrix and $\left(\Sigma_1^2\ U_1\right)$ are the p dominant components of the matrix $\left(H\,R_-^{-1}\,H^*\right)$.

$$H R_-^{-1} H^* = [U_1 \; U_2] \begin{bmatrix} \Sigma_1^2 & 0 \\ 0 & \Sigma_2^2 \end{bmatrix} \begin{bmatrix} U_1^* \\ U_2^* \end{bmatrix}$$

5.4.2 - The canonical correlation criterion [35]

This criterion relies upon the mutual information [36] which is useful to extract the most predictive part of an observed process.

The mutual information between two random vectors is the amount of information about one vector, contained in the other one, and, under gaussian hypothesis, can be simply expressed by means of the canonical correlation coefficients [37], [38] between both vectors.

For the array output, the mutual information between, the future and the past observations is given by:

$$I(Y_+, Y_-) = -\sum_k \log\left(1 - \sigma_k^2\right)$$

where the canonical correlation coefficients $\{\sigma_k\}$, $1 > \sigma_k > 0$, are defined as the non-zero singular values of the cross-covariance matrix of the future and past normalized observation vectors:

$$R_+^{-1/2} H \left(R_-^{-1/2}\right)^* = E\left[\left(R_+^{-1/2} Y_+\right)\left(R_-^{-1/2} Y_-\right)^*\right]$$

Obviously, the most important part of the information for a given order p process, is obtained by retaining the p largest canonical variables.

This is achieved by looking for a p x 1 state vector X belonging to the space of the past observations, $X = A Y_-$, and maximizing the mutual information with the future observations:

$$\underset{X = A Y_-}{\text{Max}} \; I(Y_+, X) \tag{5.22}$$

with $I(Y_+, X) = H(Y_+) + H(X) - H\begin{pmatrix} Y_+ \\ X \end{pmatrix}$

where $H(Y_+)$, $H(X)$ are the uncertainties (or entropies) in Y_+, X and $H\begin{pmatrix} Y_+ \\ X \end{pmatrix}$ is their joint uncertainty.

Using the relation (5.18) and assuming the random vectors are gaussian, it can be shown that the estimation of X amounts to determine the p x n_2 matrix A which minimizes the following criterion:

$$\underset{A}{\text{Min}} \left\{\det\left[R_+ - H A^* \left(A R_- A^*\right)^{-1} A H^*\right]\right\} \tag{5.23}$$

Forgetting the special structure of A, the solution of this problem is obtained by elementary algebra [9], [11]:

$$A = B \, V_1^* \, R_-^{-1/2} \tag{5.24}$$

where B is any p x p invertible matrix and V_1 is the matrix formed from the p right principal singular vectors of the matrix $\left[R_+^{-1/2} H \left(R_-^{-1/2}\right)^*\right]$.

$$R_+^{-1/2} H \left(R_-^{-1/2}\right)^* = [U_1 \; U_2] \begin{bmatrix} \Sigma_1^2 & 0 \\ 0 & \Sigma_2^2 \end{bmatrix} \begin{bmatrix} V_1^* \\ V_2^* \end{bmatrix}$$

5.4.3 - Estimation of the bearings

Each criterion provides the best estimation, according to its own meaning, of the matrix A, and therefore, of the state vector X of a minimum phase innovation model (5.5) of given order p, and, whose transition matrix F contains the interesting information about the sources (bearings).

This matrix can be deduced either from the observability matrix of the system $\mathcal{O}$ since:

$$\begin{cases} Y_+ \mid Y_- = \mathcal{O} A Y_- \\ Y_+ \mid Y_- = H \Gamma_-^{-1} Y_- \end{cases} \quad \text{and} \quad \mathcal{O}^{\uparrow} F = \mathcal{O}^{\downarrow}$$

or from the state dynamics: $X_{i+1} \mid X_i = F_- X_i$

This last procedure leads to a stable solution (poles inside the unit circle) for both criteria [9]:

$$F_{pe} = B\, \Sigma_1\, U_1^* \left(H^{-1}\right)^* \overleftarrow{R}_-\; H^{-1}\, U_1\, \Sigma_1\, B^{-1} \tag{5.25}$$

$$F_{cc} = B\, V_1^*\, R_-^{-1/2}\, \overleftarrow{R}_-\, \left(R_-^{-1/2}\right)^*\, V_1\, B^{-1} \tag{5.26}$$

where the subscripts "pe" and "cc" stand respectively for the predictive efficiency criterion solution and the canonical correlation criterion solution, $\overleftarrow{R}_-$ is the matrix deduced from the past observation covariance matrix R by shifting to the left of one column the columns of the matrix R_- (the other matrices corresponding to the definitions previously given for each criterion).

For each of the expressions (5.25) or (5.26), the eigenvalues of the transition matrix can be derived by choosing the suitable matrix B, and hence, the source bearings are straightforwardly obtained from the phases of the pole estimates (eigenvalues of F) which are close to the unit circle.

5.4.4 - Concluding remarks

Some further results can be obtained with the predictive efficiency criterion. First of all, a link with a classical array processing expression can be easily derived in the additive white noise case [9]: $H R_-^{-1} H^* = \mathcal{D} \Delta \mathcal{D}^*$, where $\mathcal{D}$ is the $n_2 \times m$ steering matrix and Δ is a $m \times m$ matrix. So the range $\left(H R_-^{-1} H^*\right)$ is related to the source subspace.

Elsewhere, in the general case, the matrix H is only perturbed by an upper left triangular matrix under the assumption of a shortly correlated additive noise, whereas the matrix R_- is perturbed by a diagonal band matrix.

Thereby, the vectors of {Range $(H^* H) \cap$ Range (R_-)} are mainly related to the source steering vectors, and it is interesting to seek a basis of this subspace intersection. The predictive efficiency criterion provides a matrix A whose rows approximate a basis of {Range $(H^* H) \cap$ Range (R_-)}, which means that the rows of A minimize the principal angles between Range $(H^* H)$ and Range (R_-) [11]. Thus the predictive efficiency criterion can be expressed in terms of principal angles between two linear subspaces [43].

It is important to note that both criteria are based on the same matrix $\left[R_+ - H A^* \left(A R_- A^*\right)^{-1} A H^*\right]$ and differ only by the functional that they use: trace or determinant. The minimization of each of them is achieved by means of a principal component analysis and does not allow to take into consideration the special structure of the matrix A.

Nevertheless, the approximated realization methods enhance considerably the robustness w.r.t. noise correlation relatively to usual HR methods. Spurious sources due to the bumps of the noise spatial density are cancelled, and the detection of weak sources corresponding spatially to the holes of this density is improved (commonly till 3 dB, depending on the configuration).

These improvements are mainly due to the modelling which is used. Another worthwhile step is possible by taking into consideration the problem structure to obtain the reduced model. This is the intended effect of the optimal realization method which is presented in the following section

5.5 - OPTIMAL REALIZATION METHOD

In the previous approach, a low order model is directly estimated from the sensor output covariances. Other methods could be applied to solve this problem, leading to slightly different realizations (F, T, h). On the other hand, high order models are quite robust w.r.t. the identification method which is used and easy to perform.

So, the key idea is to find a maximum order (n) model well suited to describe the sensor outputs, then to reduce it in an "optimal" way in order to obtain the intended given order (p) model.

Recall that we are interested in a state space modelling of the sensor outputs, which means that the maximum order model has to be represented by an innovation model (5.5) where $\{w_i\}$ and $\{y_i\}$ are the input and output processes. A special model has to be chosen for playing the role of initial maximum order model. In our context, the maximum entropy AR model will be chosen for several reasons. This choice will be discussed further since the general method which is presented now does not require this model restriction.

Under the assumption of zero initial conditions, the z-transforms of the input-output processes of the innovation model (5.5) are connected by means of the transfer function of the linear system [30]: $\frac{y(z)}{w(z)} = f_n(z)$

With the state-space notations (5.5), the transfer function is given by:

$$f_n(z) = h^*(zI - F)^{-1}\,T + 1 \tag{5.27}$$

The z-function is rational: $$f_n(z) = \frac{n(z)}{d(z)} + 1$$

The poles of the model (n poles) are the roots of the denominator polynomial d (z), and the zeros are the roots of the numerator polynomial (n (z) + d (z)). Thanks to the stability hypothesis, the function $f_n(z)$ is analytic outside the unit circle on the z-plane (all the n poles are within the unit circle), and $f_n(z)$ has a Laurent series expansion on that domain:

$$f_n(z) = 1 + \sum_{k=1}^{\infty} c_k z^{-k} \tag{5.28}$$

where $c_k = h^* F^{k-1}\,T \qquad k \geq 1$

The $\{c_k\}$ are defined as the Markov parameters and are equal to the values of the impulse response of the discrete system, still further, c_k is the k-th Fourier coefficient of f_n w.r.t. the complete basis $\{z^{-k}\}$, $z = \exp[j\,\omega]$.

An important matrix connected with the Markov parameters is the Hankel matrix of the system:

$$H_{f_n} = \begin{bmatrix} c_1 & c_2 & c_3 & \cdots \\ c_2 & c_3 & \cdots & \\ c_3 & \cdots & & \\ \cdots & & & \end{bmatrix} = \mathcal{O}\,\mathcal{C} \tag{5.29}$$

where $\mathcal{O}$ and $\mathcal{C}$ are the infinite observability and reachability matrices:

$$\mathcal{O} = \begin{pmatrix} h^* \\ h^* F \\ h^* F^2 \\ \vdots \end{pmatrix}, \quad \mathcal{C} = (T, F\,T, F^2\,T, \dots)$$

It is well-known that the singular values of the infinite Hankel matrix provide us with information about the order of the system since they are equal to the square root of the eigenvalues of the matrix ($G_{\mathcal{C}}\, G_{\mathcal{O}}$), where $G_{\mathcal{C}} = \mathcal{C}\,\mathcal{C}^*$ and $G_{\mathcal{O}} = \mathcal{O}^*\,\mathcal{O}$ are the (finite) reachability and observability gramians, respectively. These ones are called the singular values of the system, and the number of nonzero singular values of an "exact" (not perturbed) Hankel matrix (i.e. the rank of the Hankel matrix) determines the order of a minimal realization of the observed process.

Therefore, it is important for the solution of the model reduction problem, to use the singular values of the system as a measure of the distance between the initial model and the reduced one, and the problem can be stated in the following terms: let H_{f_n} be a given Hankel matrix, find a rank p Hankel matrix H_{f_p} which best approximates the initial matrix: $\underset{f}{\mathrm{Min}} \| H_{f_n} - H_{f_p} \|_s$ (5.30)

where $\| \ \|_s$ stands for the spectral norm (largest singular value).

Relaxing the Hankel form constraint about the approximating matrix, it can be shown that this bound is achieved by means of the singular value decomposition (SVD) of the matrix H_{f_n}: $H_{f_n} = U\,\Sigma\,V^*$

where $\Sigma = \mathrm{diag}\,(\sigma_i)$, the σ_i's being positive real and in descending order.

Then:

$$\mathrm{Min} \| H_{f_n} - M \|_s = \sigma_{p+1} \tag{5.31}$$

where $M = U\,\hat{\Sigma}\,V^*$ and $\hat{\Sigma} = \mathrm{diag}\,(\hat{\sigma}_i)$ with $\hat{\sigma}_1 = \sigma_1, \dots, \hat{\sigma}_p = \sigma_p, \hat{\sigma}_{p+1} = \hat{\sigma}_{p+2} = \dots = 0$

Unfortunately, the matrix M is not a Hankel matrix in general, and therefore, there is not a corresponding rational model (no exact realization of finite order).

In this sense this approximation is not optimal. However, for a single input - single output (SISO) linear system, Adamjan, Arov and Krein showed that among the matrices which achieve this bound there is a Hankel one, and they provided an explicit solution to the infinite Hankel matrix approximation.

5.5.1 - Infinite Hankel matrix approximation

Adamjan, Arov and Krein proved the following fundamental theorem [39]:

Theorem : for an arbitrary rank n Hankel matrix H_{f_n}, there exists a rank p Hankel matrix H_{f_p} such that:

$$\| H_{f_n} - H_{f_p} \|_s = \sigma_{p+1} \tag{5.32}$$

where σ_{p+1} is the (p + 1)-th singular value of H_{f_n}.

The proof of this theorem relies upon the approximation of a rational function on the unit circle [40] which can be introduced by means of the well-known Kronecker theorem:

Kronecker's theorem : let $\psi \in L_\infty$, then H_ψ has finite rank $\leq k$ if and only if $\psi \in H_{\infty,k}$. L_∞ denotes the Banach space of essentially bounded function on the unit circle with essential supremum norm:

$$\| \psi \|_\infty = \underset{|z|=1}{\mathrm{ess\ sup}} |\psi(z)|$$

$H_{\infty,k}$ denotes the set of functions $\psi \in L_\infty$ having at most k poles within the unit circle:

$$\psi(z) = \frac{g(z)}{(z-\alpha_1)\cdots(z-\alpha_k)}$$

where $|\alpha_1| < 1, \ldots, |\alpha_k| < 1$, and $g \in H_\infty$. H_∞ denotes the subspace of essentially bounded functions (L_∞) which are analytic inside the unit circle.

Let $\psi \in L_\infty$,
$$\psi(z) = \sum_{-\infty}^{+\infty} c_i z^{-i}$$

Then ψ can be partitioned as: $\quad \psi(z) = [\psi(z)]_+ + [\psi(z)]_-$

where $[\psi(z)]_+ = \sum_{i=-\infty}^{0} c_i z^{-i}$ is the proper instable part of $\psi(z)$ $\left([\psi(z)]_+ \in H_\infty\right)$, and,

$[\psi(z)]_- = \sum_{i=1}^{+\infty} c_i z^{-i}$ is the strictly proper stable part of $\psi(z)$ $\left([\psi(z)]_- \in L_\infty \ominus H_\infty\right)$

Obviously, $H_\psi = H_{[\psi]_-}$, since the Hankel matrix is formed with the positive Fourier coefficients.

Moreover, H_ψ has rank k when $[\psi(z)]_-$ has k stable poles ($\alpha_1, \alpha_2, \ldots, \alpha_k$).

Hence, a rank k Hankel matrix H_{ψ_k} determines a unique rational transfer function $\varphi_k(z) = [\psi_k(z)]_-$ of degree k (= degree of the denominator polynomial) which is stable (all the k poles are within the unit circle) and strictly proper (degree of the numerator polynomial < k).

This leads to connect the Hankel matrix approximation problem to a transfer function approximation problem: let $\varphi_n(z)$ be a given stable strictly proper transfer function of degree n, find a stable strictly proper transfer function $\varphi_p(z)$ of degree $p < n$ which minimizes: $\quad \text{Inf}\,\|\varphi_n(z) - \varphi(z)\|_H \quad$ (5.33)
where $\|\ \|_H$ is the Hankel norm: $\|\varphi(z)\|_H = \|H_\varphi\|_s$

The preceding considerations pointed out the links between the preserving structure problems for Hankel matrices and rational models. It is now interesting to solve the basic problem of rational transfer function approximation.

In this way, the Nehari's theorem is helpful:

Nehari's theorem : let $\psi(z) \in L_\infty$. Then:

$$\|H_\psi\|_s = \|\psi(z)\|_H = \inf_{\eta \in H_\infty} \|\psi(z) - \eta(z)\|_\infty \tag{5.34}$$

Then the solution of the stable rational transfer function approximation problem is connected to the solution of the following problem:

Problem : let $f_n(z) \in L_\infty$ (f_n has degree n) and p a given order ($n > p \geq 0$). Find the function $\psi_p(z) \in H_{\infty,p}$ which minimizes: $\quad \text{dist}_{L_\infty}(f_n, H_{\infty,p}) = \text{Inf}\,\|f_n - \psi_p\|_{L_\infty} \quad$ (5.35)

The theorem of Adamjan, Arov and Krein [39] provides an explicit solution to the above problem:

Theorem (Adamjan, Arov and Krein): let $f_n \in L_\infty$ (f_n has degree n) and p a given order ($n > p \geq 0$). Then:

$$\text{dist}_{L_\infty}(f_n, H_{\infty,p}) = \sigma_{p+1} \tag{5.36}$$

where the $\{\sigma_i\}$ are the singular values of H_{f_n} arranged in descending order; furthermore the distance is attained at a unique function $\psi_p \in H_{\infty, p}$, and if (v_{p+1}, u_{p+1}) is the Schmidt pair of H_{f_n} corresponding to σ_{p+1} then:

$$\psi_p(z) = f_n(z) - \sigma_{p+1} \frac{u_{p+1}(z)}{v_{p+1}(z)} \tag{5.37}$$

where $H_{f_n} v_{p+1} = \sigma_{p+1} u_{p+1}$

$$v_{p+1}(z) = \sum_{i=1}^{+\infty} v_i^{p+1} z^{i-1}$$

$$u_{p+1}(z) = \frac{1}{\sigma_{p+1}} \left[f_n(z)\, v_{p+1}(z)\right]_- = \sum_{i=1}^{+\infty} u_i^{p+1} z^{-i}$$

and v_i^{p+1} , u_i^{p+1} denote the i-th components of v_{p+1} and u_{p+1}, respectively

$$\left(v_{p+1}(z) \in H_\infty,\ u_{p+1}(z) \in L_\infty \ominus H_\infty\right)$$

It is important to note that due to Hankel operator properties, the approximating transfer function $\varphi_p(z) = [\psi_p(z)]_-$ is strictly proper and stable, and is the approximation of the strictly proper stable part of the initial transfer function $\varphi_n(z) = [f_n(z)]_-$.

In practice, the model reduction procedure is applied to the strictly proper part of the initial (stable proper) transfer function, and then the direct term c_0 (coefficient of z^0 in the Laurent series expansion of f_n) is added to the reduced order function to obtain the intended proper stable reduced order transfer function f_p.

The preceding theorem provides a quite interesting result but which is not straightforwardly applicable because it involves the infinite Hankel matrix H_{f_n} and its singular vectors. Using linear system properties (balancing transform), a practical algorithm can be derived.

5.5.2 - Constructive solution with finite structure

Let $f_n(z)$ be the initial (stable proper) transfer function of order n, which means that it exists a rank n realization (F, T, h) such that: $f_n(z) = h^*(zI - F)^{-1} T + 1 = \frac{n(z)}{d(z)} + 1$ (5.38)

Then, its strictly proper part is given by: $\varphi_n(z) = [f_n(z)]_- = h^*(zI - F)^{-1} T = \frac{n(z)}{d(z)}$ (5.39)

We seek for a rank p approximation of this function. Following the theorem of Adamjan, Arov and Krein, the singular value decomposition of the infinite Hankel matrix H_{φ_n} $(= H_{f_n})$ is required. Fortunately, it has been see that the Hankel matrix of the system is given by: $H_{\varphi_n} = \mathcal{O}\,\mathcal{C}$ (5.40)

Due to the rank n assumption, only n singular values are non zero, and the decomposition of the infinite Hankel matrix can be written as: $H_{\varphi_n} = U \Sigma V^*$ (5.41)

where U, Σ and V are $(\infty \times n)$, $(n \times n)$ and $(\infty \times n)$ matrices, respectively ($\Sigma = \mathrm{diag}(\sigma_i)$, $\sigma_1 \geq \sigma_2 \geq \ldots \geq \sigma_n > 0$).

Then, in balanced coordinates, the reachability and observability matrices are given by:

$$\mathcal{C}_b = \Sigma^{1/2} V^*,\ \mathcal{O}_b = U \Sigma^{1/2} \tag{5.42}$$

The corresponding balanced triple is deduced from the initial realization by a balancing transform:

$$(F, T, h) \rightarrow \left(F_b = B F B^{-1}, \ T_b = B T, h_b = (B^{-1})^* h\right)$$

where B is the balancing transform matrix.

Note that (F, T, h) and (F_b, T_b, h_b) are both realizations of the same transfer function, which means:

$$\varphi_n(z) = h_b^* (z I - F_b)^{-1} T_b = \frac{n(z)}{d(z)} \tag{5.43}$$

Using the balanced realization, the Schmidt pair of H_{φ_n} corresponding to σ_{p+1} is given by:

$$\begin{cases} u_{p+1} = \mathcal{O}_b \Sigma^{-1/2} e_{p+1} \\ v_{p+1} = \mathcal{C}_b^* \Sigma^{-1/2} e_{p+1} \end{cases} \tag{5.44}$$

where e_{p+1} is the (p + 1)-th vector of the identity matrix.

Hence, the corresponding z-function are easily deduced:

$$\begin{cases} u(z) = \sum_{i=1}^{+\infty} \left(h_b^* F_b^{i-1} \Sigma^{-1/2} e_{p+1}\right) z^{-i} \\ v(z) = \sum_{i=1}^{+\infty} \left(T_b^* (F_b^*)^{i-1} \Sigma^{-1/2} e_{p+1}\right) z^{i-1} \end{cases} \tag{5.45}$$

(for more convenience, $u_{p+1}(z) = u(z)$ and $v_{p+1}(z) = v(z)$)

Then:

$$\begin{cases} u(z) = \sigma_{p+1}^{-1/2} h_b^* (z I - F_b)^{-1} e_{p+1} \\ v(z) = \sigma_{p+1}^{-1/2} T_b^* (I - z F_b^*)^{-1} e_{p+1} \end{cases} \tag{5.46}$$

Obviously, these functions are rational and strictly proper. Let u (z) be given by :

$$u(z) = \frac{m(z)}{d(z)} \tag{5.47}$$

with deg (d (z)) = n and deg (m (z)) < n. The rational function u (z) has the same denominator polynomial as $\varphi_n(z)$ because: det (z I - F_b) = det (z I - F).

Then, by means of a balanced system property [10], [42], it can be shown that the rational function v (z) is related to u (z).

Property : *for a balanced SISO system, the following equalities are satisfied:*

$$F_b^T = Q F_b Q^* \ ; \ T_b = Q^* \bar{h}_b$$

Q being a unitary diagonal matrix ($Q = diag(q_i)$, $|q_i| = 1$).

Then the above property and (5.46) yield the following fundamental property:

$$v(z) = q_{p+1} z^{-1} u^*(z^{-1}) = q_{p+1} \frac{\tilde{m}(z)}{\tilde{d}(z)} \tag{5.48}$$

where q_{p+1} is the (p + 1)-th diagonal element of the transform matrix Q defined in the preceding property, and $\tilde{m}(z)$, $\tilde{d}(z)$ are respectively the reversed polynomials of m (z) and d (z):

$$\begin{cases} \tilde{m}(z) = z^{n-1}\, m^*(z^{-1}) \\ \tilde{d}(z) = z^{n}\quad d^*(z^{-1}) \end{cases} \tag{5.49}$$

Therefore, using the theorem of Adamjan, Arov and Krein, the best L_∞- approximation of order p is derived:

$$\Psi_p(z) = \frac{n(z)}{d(z)} - \lambda \frac{m(z)\, \tilde{d}(z)}{d(z)\, \tilde{m}(z)} \tag{5.50}$$

where $\lambda = q_{p+1}^{-1}\ \sigma_{p+1}$

This equation can be written in a polynomial form as:

$$\frac{n(z)}{d(z)}\tilde{m}(z) - \lambda \frac{\tilde{d}(z)}{d(z)} m(z) = p(z) \tag{5.51}$$

where it can be easily proved that p (z) is a polynomial of degree at most (n - 1).

Indeed, the Hankel operator yields the following equality:

$$[\varphi_n(z)\ v(z)]_- = \sigma_{p+1}\ u(z) \tag{5.52}$$

or

$$[\varphi_n(z)\ v(z) - \sigma_{p+1}\ u(z)]_- = 0 \tag{5.53}$$

Substituting φ_n (z), v (z) and u (z) for their expressions, one obtains:

$$\left[\frac{n(z)\ \tilde{m}(z) - \lambda \tilde{d}(z)\ m(z)}{d(z)\ \tilde{d}(z)}\right]_- = 0 \tag{5.54}$$

The partial expansion of the above fraction is given by:

$$\frac{n(z)\ \tilde{m}(z) - \lambda \tilde{d}(z)\ m(z)}{d(z)\ \tilde{d}(z)} = \frac{p(z)}{\tilde{d}(z)} + \frac{q(z)}{d(z)} \tag{5.55}$$

where q (z) has to be zero in order to satisfy the equation (5.54) since the roots of $\tilde{d}$ (z) are outside the unit circle. Therefore, p (z) is a polynomial, and it can be easily checked from the equation (5.51) that its degree is at most (n - 1).

The best L_∞- approximation of order p is then given by:

$$\Psi_p(z) = \frac{p(z)}{\tilde{m}(z)} \tag{5.56}$$

where p (z) and m (z) satisfy the equation (5.51).

In order to obtain explicitly these polynomials, the equation (5.51) has to be written as:

$$p(z)\ d(z) = n(z)\ \tilde{m}(z) - \lambda \tilde{d}(z)\ m(z) \tag{5.57}$$

Expanding this equation and equating the coefficients of powers of z, the two following matricial equations are obtained [41] (the first one for $\{z^i\}_{i=0,n-1}$, and the second one for $\{z^i\}_{i=n,2n-1}$):

$$J\, S_2\, J\, p = R_2\, \overline{m} - \lambda\, J\, \overline{S}_1\, J\, m \tag{5.58}$$

$$S_1\, p = R_1\, \overline{m} - \lambda\ \overline{S}_2\, m \tag{5.59}$$

where:

$n(z) = c_n + c_{n-1} z + \dots + c_1 z^{n-1}$ $\qquad$ $d(z) = a_n + a_{n-1} z + \dots + a_1 z^{n-1} + z^n$

$m(z) = m_n + m_{n-1} z + \dots + m_1 z^{n-1}$ $\qquad$ $p(z) = p_n + p_{n-1} z + \dots + p_1 z^{n-1}$

$m^T = (m_1, m_2, \dots, m_n)$ $\qquad$ $p^T = (p_1, p_2, \dots, p_n)$

$$S_1 = \begin{bmatrix} a_{n-1} & a_{n-2} & \cdots & 1 \\ a_{n-2} & & \cdot^{\cdot^{\cdot}} & \vdots \\ \vdots & \cdot^{\cdot^{\cdot}} & & \vdots \\ 1 & \cdots & \cdots & 0 \end{bmatrix} \qquad S_2 = \begin{bmatrix} a_1 & a_2 & \cdots & a_n \\ a_2 & & \cdot^{\cdot^{\cdot}} & \vdots \\ \vdots & \cdot^{\cdot^{\cdot}} & & \vdots \\ a_n & \cdots & \cdots & 0 \end{bmatrix}$$

$$R_1 = \begin{bmatrix} 0 & c_1 & \cdots & c_{n-1} \\ \vdots & & \ddots & \vdots \\ \vdots & & & c_1 \\ 0 & \cdots & \cdots & 0 \end{bmatrix} \qquad R_2 = \begin{bmatrix} c_n & \cdots & \cdots & 0 \\ \vdots & \ddots & & \vdots \\ c_2 & & \ddots & \vdots \\ c_1 & c_2 & \cdots & c_n \end{bmatrix}$$

$$J = \begin{bmatrix} 0 & \cdots & 1 \\ \vdots & \cdot^{\cdot^{\cdot}} & \vdots \\ 1 & \cdots & 0 \end{bmatrix}$$ (J = antidiagonal unity matrix)

Substituting p, a pseudo generalized eigen-equation is derived:

$$\left(J\, S_2\, J\, S_1^{-1} R_1 - R_2\right)\overline{m} = \lambda \left(J\, S_2\, J\, S_1^{-1}\, \overline{S}_2 - J\, \overline{S}_1\, J\right) m \qquad (5.60)$$

where $\lambda = q_{p+1}^{-1}\ \sigma_{p+1}$.

Noting that $|q_{p+1}| = 1$, the complex scalar λ can be written: $\lambda = e^{j\eta}\sigma_{p+1}$ and the eigen-equation becomes:

$$\left(J\, S_2\, J\, S_1^{-1} R_1 - R_2\right)\overline{m}_\eta = \sigma_{p+1} \left(J\, S_2\, J\, S_1^{-1}\, \overline{S}_2 - J\, \overline{S}_1\, J\right) m_\eta \qquad (5.61)$$

where $m_\eta = e^{j\eta/2}\, m$ and σ_{p+1} is a real scalar.

Decomposing into real and imaginary components the matrices as follows:

$$J\, S_2\, J\, S_1^{-1} R_1 - R_2 = A_r + j\, A_j$$
$$J\, S_2\, J\, S_1^{-1}\, \overline{S}_2 - J\, \overline{S}_1\, J = B_r + j\, B_j$$
$$m_\eta = m_r + j\, m_j$$

the complex valued n x n eigenrelationship is transformed into the equivalent 2n x 2n real valued eigenrelationship

$$\begin{pmatrix} A_r & A_j \\ A_j & -A_r \end{pmatrix} \begin{pmatrix} m_r \\ m_j \end{pmatrix} = \sigma_{p+1} \begin{pmatrix} B_r & -B_j \\ B_j & B_r \end{pmatrix} \begin{pmatrix} m_r \\ m_j \end{pmatrix} \qquad (5.62)$$

and the selection of the eigenvector corresponding to the (p + 1)-th eigenvalue σ_{p+1} (the eigenvalues being in descending order) provides the intended vector m_η.

Remark : it is clear that both pairs (σ_{p+1}, m_η) and $(-\sigma_{p+1}, j\, m_\eta)$ are solutions of the generalized eigenproblem.

Then, from the eigensolution of (5.62) and (5.59), it can be easily deduced:

$$p_\eta = S_1^{-1}\left(R_1\, \overline{m}_\eta - \sigma_{p+1}\, \overline{S}_2\, m_\eta\right) \qquad (5.63)$$

where $p_\eta = e^{-j\eta/2}\, p$.

Obviously the pair (p_η, m_η) satisfies the following equality:

$$\frac{p_\eta(z)}{\tilde{m}_\eta(z)} = \frac{p(z)}{\tilde{m}(z)} \tag{5.64}$$

Therefore, an explicit solution of the theorem of Adamjan, Arov and Krein is derived using the equations (5.62) and (5.63) as:

$$\psi_p(z) = \frac{p_\eta(z)}{\tilde{m}_\eta(z)} \tag{5.65}$$

Expanding this fraction to obtain the strictly proper stable part of $\psi_p(z)$, $\varphi_p(z) = [\psi_p(z)]_-$, and adding the coefficient of z^0 in the Laurent series expansion of f_n, the order p rational approximation is obtained:

$$f_p(z) = 1 + \left[\frac{p_\eta(z)}{\tilde{m}_\eta(z)}\right]_- \tag{5.66}$$

This general solution can be rather simplified according to the initial transfer function which is given to reduce.

5.5.3 - Choice of the initial model and its implications

The preceding model reduction procedure is applied to an initial maximum order model which has to be estimated from the sensor output data. As it has been already said, high order models are quite robust w.r.t. the identification method which is used; nevertheless, this step is of great importance.

In our context, the most random extension, i.e. the maximum entropy extension seems to be the most natural and careful extension. For an equally spaced line array, this leads to an autoregressive (AR) parametrization of order n for (n + 1) sensors. Note that there is no information loss since it is simply another representation of the given covariances. So, an AR modelling can be deduced from the partial sequence of covariance estimates, and the model reduction procedure can be applied to the following strictly proper stable function $\varphi_n(z)$ which is a spectral factor (on the unit circle) of the spatial density of the estimated model:

$$\varphi_n(z) = \frac{1}{z^n + a_1 z^{n-1} + \ldots + a_n} \tag{5.67}$$

The use of this special rational function and the fact that the poles are the only interesting parameters for us lead to simplify the model reduction algorithm [10], [42].

First of all, it is important to note that $\varphi_n(z)$ belongs to L_∞, since the AR estimation provides a "strictly" stable model . However, the poles corresponding to plane waves lie close to the unit circle and hence have great importance for the L_∞-norm, whereas weak sources can have small contributions in the meaning of the L_2-norm.

Let $\psi_p(z)$ be the L_∞ approximation of order p of $\varphi_n(z)$: $\psi_p(z) = \dfrac{p_\eta(z)}{\tilde{m}_\eta(z)}$

As the poles are the only interesting parameters, it is useless to compute the numerator polynomial.

As for the denominator polynomial, its computation is rather simple. Indeed, with the given $\varphi_n(z)$, $m_\eta(z)$ is obtained by solving a "simple" eigenrelationship instead of a generalized one:

$$\begin{pmatrix} -B_r & B_j \\ B_j & B_r \end{pmatrix} \begin{pmatrix} m_r \\ m_j \end{pmatrix} = \sigma_{p+1}^{-1} \begin{pmatrix} m_r \\ m_j \end{pmatrix} \tag{5.68}$$

since $R_1 = 0$ and $R_2 = I$, here (the notations being defined above).

Then, the source bearing estimates are easily deduced. The practical implementation of this array processing method can be summarized in the following algorithm outline:

1) estimation of the sensor output covariances (n + 1 sensors); i.e.: $\hat{r}_0, \hat{r}_1, \ldots, \hat{r}_n$;
2) estimation of the corresponding AR model coefficients: $(1, \hat{a}_1, \ldots, \hat{a}_n)$;
3) solve the eigen-problem (5.68) with $a_1 = \hat{a}_1, \ldots, a_n = \hat{a}_n$; select the pertinent eigenvector (corresponding to σ_{p+1} in order to obtain a rank p approximation) and form the polynomial $\tilde{m}_\eta(z)$ whose coefficients are straightforwardly deduced from the elements of the preceding vector;
4) compute the roots of $\tilde{m}_\eta(z)$, and the arguments of the p roots having a modulus less than unity (stable part of $\psi_p(z)$) provide the source bearing estimates. Note that by definition of $H_{\infty, p}$, this polynomial has exactly p roots within the unit circle.

5.5.4 - Simulation results

The performances of the optimal realization method (ORM) are compared with those of the root MUSIC method. For that purpose an equally spaced line array constituted with 20 sensors (at $\lambda/2$) is considered. The broadside corresponds to 90°.

Simulation 1 : the additive noise is spatially white, and 3 sources are simulated with respective bearing and SNR (30°, - 10 dB), (36°, - 20 dB) and (45°, - 3 dB).

The number of averaged snapshots (BT) is equal to 200. The results of the detection of the weak source at 36° are given for 10 independent trials in the following table:

Detection of the weak source (36°)	Root-MUSIC	ORM
	0/10	8/10

Table 1

These results show that the root-MUSIC method is unable to detect the weak source, whereas the ORM detects it 8 times on 10 trials.

Simulation 2 : the additive noise is now strongly correlated and simulated by means of an AR model of order 6. Its spatial density corresponds to a surface noise: figure 11.

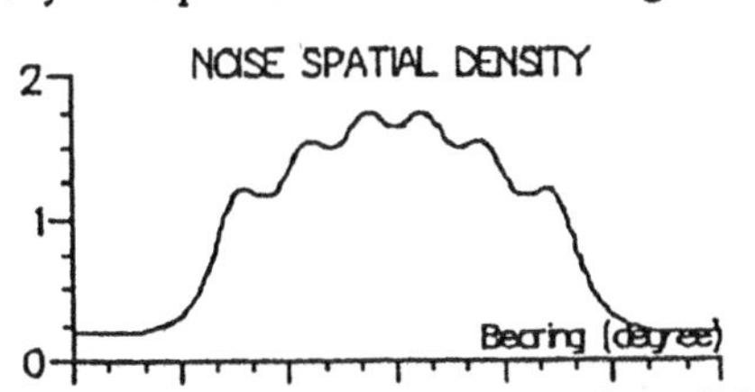

Figure 11: spatial density of the simulated noise

3 sources are simulated with respective bearing and power (10°, 3), (17°, 0.3) and (70°, 0.03). The number of averaged snapshots is equal to 200. The analysis of both method results indicates that the very weak source (70°) corresponding spatially to the bump of the noise spatial density is well detected, whereas

the weak source (17°) located in a hole of the noise spatial density is difficult to detect. For 10 independent trials, the detection rates of this source for both methods are compared in the following table 2:

Detection of the weak source (17°)	Root-MUSIC	ORM
	0/10	7/10

Table 2

In view of these results, it is clear that the ORM is much more robust than the root-MUSIC method w.r.t. the noise spatial correlation.

5.5.5 - Multifrequency extension

The aim of this section is to present briefly an extension of the optimal realization method to the multifrequency analysis. This analysis has the advantage of providing a unique estimation of wide band source bearings by integrating the analysis at each frequency, and it is well-known that the processing performances can be improved by means of a coherent frequency integration. Therefore, it is better to analyse coherently the whole frequency band instead of summing up the narrow band processing outputs at each frequency.

The optimal realization method lends itself very well to a coherent multitrequency analysis, as it can be seen in the following.

The line array is assumed to be composed with (n + 1) equally spaced sensors.Let $[f_1, f_2]$ be the frequency band of interest.Under spatial stationarity hypothesis, the elements of the cross-spectral matrix at a given frequency $f_i \in [f_1, f_2]$ are characterized by the frequency f_i and the distance (q d) between the corresponding pairs of sensors: $r(f_i, q\,d)$ (d is the distance between two contiguous sensors, and q is an integer : $-n \leq q \leq n$).

In the narrow band case, a maximum order AR model is estimated from the covariances at a given frequency and then reduced by means of the optimal model reduction method.Here, it is intended to estimate a unique maximum order AR model from the covariances of the whole frequency band of interest, and then to reduce it as in the narrow band case.For that purpose, it is necessary to interpolate the covariances at each frequency to obtain the corresponding covariances at a reference frequency: let f_1 be this frequency of reference:

$$r(f_1, q\,d) = r\left(f_i, \frac{f_1}{f_i}\, q\,d\right) \tag{5.69}$$

This leads to substitute the elementary spatial sampling d by the sampling $\left(\frac{f_1}{f_i}\, d\right)$, and the interpolated covariances can be seen as covariances between fictitious sensors.The interpolation can be achieved by means of an AR modelling and the use of the well-known Wiener-Kinchine's theorem.Indeed, an order n AR model can be estimated at each frequency $f_i \in \,]f_1, f_2]$ from the covariances $\{r(f_i, q\,d), q = 0, 1, \ldots, n\}$, and then the spatial density of this AR-process $P(f_i, k)$ can be deduced as a continuous function of the wavenumber (spatial frequency) k:

$$P(f_i, k) = \frac{\sigma_i^2\, d}{\left|1 + \sum_{s=1}^{n} a_s \exp[-j\, 2\pi\, s\, k\, d]\right|^2} \tag{5.70}$$

where $\{a_s\}_{s=1,n}$ are the estimated AR coefficients and σ_i^2 is the variance of the driving white noise.

And using the Wiener-Kinchine's theorem, the interpolated covariances are derived:

$$r\left(f_i, \frac{f_1}{f_i} q d\right) = \int_{-\frac{W}{2}}^{\frac{W}{2}} P(f_i, k) \exp\left[-j 2 \pi k \frac{f_1}{f_i} q d\right] dk \qquad (5.71)$$

with $\frac{W}{2} = \frac{f_i}{f_1} \frac{1}{2d}$.

Therefore, a set of (n + 1) covariance lags $\{r(f_1, q\,d), q = 0, 1, \ldots, n\}$ can be estimated at each frequency f_i (for $f_i = f_1$, without interpolation), and a unique order n AR model can be derived from all of them in a least squares sense. Then, the model reduction of this process leads to multifrequency optimal realization method output.

This procedure can be interpreted in terms of spatial densities.

Indeed, after interpolation, the poles corresponding to sources are identical for the different AR models estimated at each frequency $f_i \in [f_1, f_2]$. Let $z_k = \exp\left[-j 2 \pi f_1 \frac{d}{c} \cos \theta_k\right]$ be the pole corresponding to the k-th source, Then, the expansion of the spatial density of the interpolated process from the frequency f_i is given by:

$$\sum_{k=1}^{m} \left(\frac{n_{k,i}}{z - z_k} + \frac{\overline{n}_{k,i}}{z^{-1} - \overline{z}_k}\right) + n_i \qquad (5.72)$$

where $n_{k,i}$ is a scalar which can be zero if the contribution of the k-th source is not enough meaningful at the frequency f_i, n_i is related to the noise (the subscript i stands for the frequency f_i), and m is the source number.

So, integrating over the frequency band $[f_1, f_2]$ and permuting both sums, the sum of the all contributions to the 2D (space-frequency) density is obtained:

$$\sum_{k=1}^{m} \left(\frac{N_k}{z - z_k} + \frac{\overline{N}_k}{z^{-1} - \overline{z}_k}\right) + N \qquad (5.73)$$

where N_k is the contribution of the k-th source to the 2D density, and N is related to the noise. The interpolation effects on the noise are not important since the optimal realization method is robust w.r.t. the noise correlation.

Hence, the integration of the interpolated process over the frequency band is justified.

5.5.6 - Concluding remarks

The optimal realization method is especially well-suited to search spatial lines in continuous noise.

Instead of estimating a low order model as the approximated realization method does, the optimal realization method works in two steps. The first one consists in estimating a high order rational model, in the circumstances a maximum entropy AR model, and is very robust thanks to the good flexibility of high order models (the choice of the spectral factor (5.67) for the reduction appears to be the best one till now).

The second step of the method allows to obtain a reduced order model having the same special structure, namely the rational structure which guarantees the existence of a state-space realization and preserves the Hankel structure. Furthermore, the model reduction is achieved using the L_∞-norm which is more sensitive to spatial line than the usual L_2-norm.

Hence, the processing improvements relatively to usual HR methods are quite meaningful: very good robustness w.r.t. the noise correlation, and improvement of the detection of weak sources in a background correlated noise and in white noise also.

Furthermore, the method can be extended to the multifrequency analysis in an easily implemented way.

Nevertheless, the problem of estimating the source number remains, although the singular values of the system can provide some informations on the system rank, and the optimal realization method can sometimes be more sensitive to the source number estimate than the usual HR methods.

6 - GENERAL CONCLUSION

The initial problem consists in extracting plane waves from the array outputs in presence of an unknown correlated noise. It is typically a measure theory problem but this aspect is occulted by algebraic calculations, especially for HR methods. The drawbacks of these methods induced by noise correlations have been explained and fully justify noise model estimation. Furthermore, the study has proved that an AR(MA) noise modelling is convenient for most of the cases.

Several methods for noise model estimation have been considered, they all rely upon parametric (ARMA) noise modelling and use no a priori assumption about it (isotropy or other...). We have tried to present them in a general way as also their main properties. Feasible and efficient algorithms for noise model estimation have been derived, they all require iterative optimization methods. Fortunately, the optimization procedure is only relative to noise parameters and the results, both on simulated and real data, are quite satisfying. After what, extraction of plane waves can be achieved by any "classical" (more or less sophisticated) method of array processing.

The other side consists in using state space modelling of the sensor outputs in order to extract directly the plane waves. Two concepts are of fundamental importance for that purpose: prediction of future by past, approximation of a transfer function (in Hilbert space). Feasible methods for array processing have been derived from these concepts, their properties have been detailed.

APPENDIX A

This appendix addresses the basic question of accuracy of noise parameters estimation in the general case. In that follows, the following hypotheses are made:

1) $\hat{R}$ is replaced by R (the exact covariance matrix of the sensor outputs).

2) The decomposition of R in source and noise part (i.e. $R = S + B_0$) is assumed to be unique.

The REF can be written as: $H = \text{Log det}\,(S + (B_0 - B)) + L \,.\, \text{Log det}\, B$

where: S is a rank s matrix (s = source number) (A.1)

The matrix B_0 being assumed to be positive definite, it can be factorized in triangular factors:

$$B_0 = T_0 \,.\, T_0^*$$

then:

$$\text{Log det}\left[S + (B_0 - B)\right] = \text{Log det}\left[S \,.\, T_0^{-1\,*} \,.\left(I_d - T_0^{-1} \,.\, B \,.\, T_0^{-1\,*}\right)^{-1} .\, T_0^{-1} + \text{Id}\right] + \text{Log det}\,(B_0 - B) \qquad \text{(A.2)}$$

Consider now, the spectral decomposition of the hermitian matrix $\left(Id - T_0^{-1} . B . T_0^{-1 *}\right)^{-1}$, i.e.:

$$\left(Id - T_0^{-1} . B . T_0^{-1 *}\right)^{-1} = W . \Delta . W^* \qquad (A.3)$$

(W: unitary matrix constituted by eigenvectors, Δ diagonal matrix of eigenvalues).

Then:
$$\Delta(i, i) = \left(1 - \lambda_i^w\right)^{-1}$$

and:
$$\text{Log det}[S + (B_0 + B)] = \text{Log det}(B_0 - B) + \text{Log det}\left[S . T_0^{-1 *} . W . \Delta . W^* . T_0^{-1} + Id\right] \qquad (A.4)$$

For the sequel, it is necessary to preserve the symmetry of the problem; for that purpose (A.4) can be symmetrized by use of the following equality:

$$\text{Log det}\left[S . T_0^{-1 *} . W . \Delta . W^* . T_0^{-1} + Id\right] = \text{Log det}\left[\Delta^{1/2} . W^* . S' . W . \Delta^{1/2} + Id\right] \qquad (A.5)$$

with:
$$S' = T_0^{-1} . S . T_0^{-1 *}$$

This last equality resulting from intensive use of the classical formula: det (A.B) = det (B.A) (valid for invertible square matrices).

$\Delta^{1/2}$ in (A.5) is the diagonal matrix defined by $\Delta^{1/2}(i, i) = (\Delta(i, i))^{1/2}$, its existence follows from the hypothesis $(B_0 - B)$ definite positive.

(A.5) may also be written as:

$$\begin{aligned}\text{Log det}\left[S . T_0^{-1} . W . \Delta . W^* . T_0^{-1} + Id\right] &= \text{Log det}\left[\Delta^{1/2} W^* . \left(S' + W . \Delta^{-1} . W^*\right) . W . \Delta^{1/2}\right] \\ &= \text{Log det}\,\Delta + \text{Log det}\left(S' + W . \Delta^{-1} . W^*\right) \qquad (A.6)\end{aligned}$$

Therefore, using classical formula for differentiation of the determinant of matrices depending on parameters λ [18] [i.e. $(\partial/\partial\lambda)$ Log det $A(\lambda) = \text{tr}\,(A^{-1}(\lambda) . (\partial/\partial\lambda) A(\lambda))$], one obtains the following expression of the derivatives of H w.r.t. parameters $\{\lambda_i^w\}$:

$$\frac{\partial}{\partial \lambda_i^w} H = L . \frac{1}{\lambda_i^w} - \frac{1}{1 - \lambda_i^w} + \left\{\frac{1}{1 - \lambda_i^w} + \text{tr}\left[\left(S' + W \Delta^{-1} . W^*\right)^{-1} . W . \frac{\partial}{\partial \lambda_i^w} \Delta^{-1} . W^*\right]\right\} \qquad (A.7)$$

Now:

$$\text{tr}\left[\left(S' + W \Delta^{-1} W^*\right)^{-1} . W . \frac{\partial}{\partial \lambda_i} \Delta^{-1} . W^*\right] = - W_i^* . \left((S' + W . \Delta^{-1} . W^*\right)^{-1} . W_i \qquad (A.8)$$

Furthermore (the notation $A \geq B$ meaning that matrix A - B is positive) :

$$S' + W . \Delta^{-1} . W^* \geq W . \Delta^{-1} . W^*$$

Whence:
$$(S' + W . \Delta^{-1} . W^*)^{-1} \leq (W . \Delta^{-1} W^*)^{-1} = W . \Delta . W^*$$

$$(A \geq B \Rightarrow A^{-1} \leq B^{-1})$$

and, finally, one obtains:

$$- W_i^* . \left(S' + W . \Delta^{-1} . W^*\right)^{-1} . W_i \geq - W_i^* . \left(W . \Delta . W^*\right) . W_i = - \frac{1}{1 - \lambda_i^w} \qquad (A.9)$$

(this last inequality becoming an equality if and only if there is no source).

In conclusion, the term between brackets in formula (A.7) is positive $\left(\lambda_i^w \in]0, 1[\right)$. Consequently, the partial derivatives of H relatively to λ_i^w are positive when λ_i^w belongs to the interval $]0, L/L+1[$.

Furthermore when λ_i^w tends towards 1^- then H tends towards $-\infty$; the REF being a concave functional (cf. prop. 3) on the convex domain $\mathscr{C}$ ($\mathscr{C}$ = { B s.t R - B and B are positive definite}) one concludes that the unique maximum of H on $\mathscr{C}$ will be achieved for $\{\beta_j\}$ values such that the corresponding $\{\lambda_i^w\}$ belong to the interval $]L/L+1, 1[$ (for i = 1, 2, ..., q).

Comments : the preceding calculations need some comments:

1) In the proof of prop. 1, the hypothesis R - B definite positive has been replaced by the hypothesis B_0 - B definite positive.

   In fact the two subsets $\mathscr{C}$ and $\mathscr{C}'$ ($\mathscr{C}'$ = { B s.t B and B_0 - B positive}) are convex and $\mathscr{C}'$ is included to $\mathscr{C}$. The functional H being concave on $\mathscr{C}$ (prop. 3) and presenting a maximum on $\mathscr{C}'$, this maximum is unique and satisfies prop. 1 on the whole subset $\mathscr{C}$.

2) As it has been seen in the proof of prop. 1, the effect of sources is to move the maximum of H (w.r.t. the $\{\lambda_i^w\}$ and to destroy the (theoretical) equality of all the λ_i^w. Thus, in presence of sources, the matrix B_0 cannot be perfectly estimated by maximization of H.

3) Property 1 is still valid when the noise model order is overdetermined, this fact follows clearly from the proof of prop. 1. For practical applications, it is a result of fundamental importance.

4) Very roughly, the functional Log det (R - B) appears as a barrier functional forbidding description of sources by means of the noise model (then the spatial coherence of sources (rank of S) plays a fundamental role.

APPENDIX B

This appendix is devoted to the determination of the optimal step of the gradient's algorithm on $\mathscr{C}$.

One wants to obtain an explicit formulation of H $(\hat{R}, B_{k+1})$ w.r.t. ρ; for that purpose, consider the Choleski decomposition of B_k and $(\hat{R} - B_k)$ in triangular factors, i.e.:

$$B_k = T_k \cdot T_k^* \quad \text{and} \quad \hat{R} - B_k = S_k \cdot S_k^* \tag{B.1}$$

Consequently:

$$\begin{aligned} \text{Log det}\left(\hat{R} - B_k + \rho\, D_k\right) &= \text{Log det}\left(S_k \cdot S_k^* + \rho\, D_k\right) \\ &= \text{Log det}\left[S_k \cdot \left(\text{Id} + \rho\, S_k^{-1} \cdot D_k \cdot S_k^{-1\,*}\right) \cdot S_k^*\right] \\ &= \text{Log det}\,(R - B_k) + \text{Log det}\left(\text{Id} + \rho\, S_k^{-1} \cdot D_k \cdot S_k^{-1\,*}\right) \end{aligned} \tag{B.2}$$

By the same manner, one obtains:

$$\text{Log det}\,(B_k - \rho\, D_k) = \text{Log det}\, B_k + \text{Log det}\left(\text{Id} - \rho \cdot T_k^{-1} \cdot D_k \cdot T_k^{-1\,*}\right) \tag{B.3}$$

Therefore, by use of (B.2) and (B.3) one obtains:

$$H(\rho) = \text{Log det}\left(\text{Id} + \rho \cdot S_k^{-1} \cdot D_k \cdot S_k^{-1\,*}\right) + L \cdot \text{Log det}\left(\text{Id} - \rho \cdot T_k^{-1} \cdot D_k \cdot T_k^{-1\,*}\right) + \text{cst} \tag{B.4}$$

The two matrices $S_k^{-1} \cdot D_k \cdot S_k^{-1\,*}$ and $T_k^{-1} \cdot D_k \cdot T_k^{-1\,*}$ are hermitian and therefore diagonalisable. Let $\{\lambda_i^k\}$ and $\{\mu_i^k\}$ their respective eigenvalues, then:

$$H(\rho) = \sum_{i=1}^{q} \text{Log}\left(1 + \rho \cdot \lambda_i^k\right) + L \cdot \sum_{i=1}^{q} \text{Log}\left(1 - \rho \cdot \mu_i^k\right) + \text{cst} \tag{B.5}$$

This last expression having the great interest to be explicit in ρ, furthermore the two constraints of positivity ($\mathscr{C}$) are translated in explicit constraints (w.r.t. ρ) i.e.:

$$1 + \rho \,.\, \lambda_i^k > 0 \qquad i = 1, 2, \ldots, q \qquad (\hat{R} - B_{k+1} \text{ positive definite})$$
$$1 - \rho \,.\, \mu_i^k > 0 \qquad i = 1, 2, \ldots, q \qquad (B_{k+1} \text{ positive definite}) \tag{B.6}$$

The optimal step ρ_k will be determined by maximization of H (ρ) (written as in (B.5)) under constraints (B.6), it is easily computed by means of an **unidimensional** Newton method starting at $\rho = 0$.

The computation of the optimal step may be of a great practical importance, especially in presence of strong sources. The main drawback of the described method is its computation cost, in [13] faster methods are given for maximization of H using special hypothesis (AR noise modelling). Note furthermore that this procedure can be extended to spatial analysis leading a very cheap estimation of the optimal step size.

APPENDIX C

This appendix deals with the proof of prop. 5, the REF H is given by formula (3.32) and its partial derivatives are given by:

$$\frac{\partial H}{\partial b_i} = - \int_{-W}^{W} \frac{\mathrm{Re}\left(z^i \,.\, \overline{F}_0(k)\right) . \left[L \,.\, R(k) - (L+1) \,.\, B(k)\right]}{\left[R(k) - B(k)\right] . B(k)} \, dk \tag{C.1}$$

Under the zero source hypothesis, we see (direct consequence of (C.1)) that the partial derivatives $\frac{\partial H}{\partial b_i}$ are null when b_i is equal to $b_i^0 \,.\, \left(\frac{L}{L+1}\right)^{1/2}$ (b_i^0 : exact value of parameter). H being a concave functional, one deduces that:

$$\hat{b}_{i,L} = b_i^0 \,.\, \left(\frac{L}{L+1}\right)^{1/2} \qquad \text{for } i = 0, 1, \ldots, p \tag{C.2}$$

In the presence of sources, things are less obvious. Let us consider the following sum of partial derivatives.

$$\sum_{i=1}^{p} b_i \,.\, \frac{\partial H}{\partial b_i}$$

then:

$$\sum_{i=1}^{p} b_i \,.\, \frac{\partial H}{\partial b_i} = \overbrace{L \,.\, \int_{-W}^{W} \frac{S(k)}{R(k) - B(k)} \, dk}^{(a)} + \overbrace{L \,.\, \int_{-W}^{W} \frac{B_0(k) - (L + 1/L) \,.\, B(k)}{R(k) - B(k)} \, dk}^{(b)} \tag{C.3}$$

(with $R(k) = S(k) + B_0(k)$, $S(k)$: source part)

The term (a) of (C.3) is positive because S (k) and R (k) - B (k) are positive (for all k) by assumption. Let us now examine the sign of (b) in (C.3).

We can assume that the (positive) term R (k) - B (k) is bounded, let be:

$$0 \leq R(k) - B(k) \leq M \quad \text{for all } k \in [-W, +W]$$

therefore:

$$(b) \geq \frac{L}{m} \,.\, \int_{-W}^{W} \left[B_0(k) - \left(\frac{L+1}{L}\right) . B(k)\right] dk = \frac{L}{M}\left(\sum_{i=0}^{p} \left\{\left| b_i^0 \right|^2 - \left(\frac{L+1}{L}\right) . \left| b_i \right|^2\right\}\right) \tag{C.4}$$

Consequently, if:

$$0 \leq |b_i| < \left(\frac{L}{L+1}\right)^{1/2} . |b_i^0| \qquad (i = 0, 1, \ldots, p)$$

then:

$$\sum_{i=0}^{p} b_i . \frac{\partial H}{\partial b_i} > 0 \tag{C.5}$$

Now, if the parameters $\{b_i\}$ approach their exact values $\{b_i^0\}$ then the REF approaches $-\infty$; furthermore using the plane wave hypothesis (calculation of term (a)) one can show that:

$$\left(\frac{L}{L+1}\right)^{1/2} . b_i^0 \leq \hat{b}_{i,L} \leq b_i^0 \tag{C.6}$$

$\left(b_i^0 \geq 0\right)$

In fact, the proof of prop. 5 is essentially similar to that of prop. 1, however the asymptotic (in n_s) expression of H avoids some tedious calculations.

REFERENCES

[1] G. Bienvenu and L. Kopp, "Principe de la goniométrie passive adaptative", VIIème colloque GRETSI, Nice, 1979, pp. 106-110.

[2] G. Bienvenu and L. Kopp, "Optimality of high resolution array processing using the eigensystem approach", IEEE Trans on ASSP, Vol. ASSP-31, n° 5, pp. 1235-1248, October 1983.

[3] L. Kopp and G. Bienvenu, "Méthodes Haute Résolution : présentation statistique", Actes de la réunion SEE du 25/04/84, Paris.

[4] G. Bienvenu and L. Kopp, "Méthodes haute résolution après formation de voies", Actes du 10ème colloque GRETSI, pp. 325-330, Nice, Mai 1985.

[5] P. Forster, "Méthodes de traitement d'antenne après filtrage spatial", Thèse de Doctorat de l'Université de Rennes I, Rennes, Novembre 1988.

[6] P. Forster and G. Vezzosi, "Application of spheroidal sequences to array processing", Proc. of the ICASSP, Dallas, pp. 2268-2271, April 1987.

[7] U. Paulraj and T. Kailath, "Eigenstructure methods for direction of arrival estimation in the presence of unknown noise fields", IEEE Trans on ASSP, Feb. 1986, Vol. ASSP-34, n° 1, pp. 13-21.

[8] S. Prasad, R.T. Williams, A.K. Mahalanabis and L.H. Sibal, "A Transform-Based Convariance Differencing Approach for Some Classes of Parameter Estimation Problem", IEEE Trans on ASSP, Vol. ASSP-36 n° 5, pp. 631-642.

[9] P. Ravazzola et J.P. Le Cadre, "Représentation des sorties de capteurs par un système linéaire, applications au traitement d'antenne", Traitement du Signal, Vol. 6, n° 1, pp. 3-26, 1989.

[10] J.P. Le Cadre and P. Ravazzola, "Approximated stochastic realization and model reduction methods applied to array processing by means of state space models", Proceedings ICASSP 1989, 23-26 May 1989, Glasgow, pp. 2601-2604.

[11] P. Ravazzola and J.P. Le Cadre, "On the use of the theory of linear systems for array processing", Submitted to the International Journal on Adaptive Control and Signal Processing.

[12] P.Y. Arquès, J.P. Le Cadre and M. Bouvet, "Les bruits en acoustique sous-marine", Actes du 11ème colloque GRETSI, pp. 1B-19B, Nice, Juin 1987.

[13] J.P. Le Cadre, "Contribution à l'utilisation des méthodes paramétriques en traitement d'antenne", Thèse de Doctorat d'Etat, Oct. 1987, USMG-INPG, Grenoble.

[14] O.S. Halpeny and D.G. Childers, "Composite Wavefronts Decomposition via Multidimensional Digital Filtering of Array Data", IEEE Trans Circuit Syst., Vol. CAS-22, pp. 552-562, June 1975.

[15] R.J. Urick, "Principles of Underwater Sound", Mc Graw Hill, 1983.

[16] J.P. Le Cadre, "High Resolution Methods in Presence of Noise Correlated Sensor Outputs. A Relative Entropy Approach, Method and Results", Submitted to IEEE Trans on ASSP.

[17] J.P. Le Cadre et P. Lopez, "Estimation d'une matrice de structure imposée. Applications", Traitement du signal, Vol. 1, n° 1, pp. 3-17, 1984.

[18] K.S. Miller, "Complex Stochastic Processes, an Introduction to Theory and Applications", Addison Wesley 1974.

[19] G.H. Golub and C.F Van loam, Matrix computations, North Oxford Academic, Oxford, 1983.

[20] U. Grenander and G. Szegö, "Toeplitz Forms and Their Applications", University of California Press, San Francisco, 1958.

[21] T.E. Barnard, "Two Maximum Entropy Beamforming Algorithms for Equally Spaced Line Arrays", IEEE Trans on ASSP, Vol. ASSP-30, n° 2, April 1982, pp. 175-188.

[22] P. Gounon, J.P. Le Cadre et P. Ravazzola, "Estimation d'un modèle du bruit reçu sur une antenne de grande dimension", 12ème colloque GRETSI, 12-16 Juin 1989, Juan-les-Pins, pp. 569-572.

[23] J.P. Le Cadre, "Parametric Methods for Spatial Signal Processing in the Presence of Unknown Colored Noise Fields", IEEE Trans on ASSP, Vol. ASSP, July 1989.

[24] T. Kailath, A. Vieira and M. Morf, "Inverses of Toeplitz Operators, Innovations and Orthogonal Polynomials", SIAM review, Vol. 20, n° 1, pp. 106-119, Jan. 1979.

[25] P. Lancaster and M. Tismenestsky, "The Theory of Matrices with applications" (second edition), Academic Press, 1985.

[26] H. Akaike, "A New Look at the Statistical Model Identification", IEEE Trans Autom. Control, Vol. AC-19, pp. 716-723, Dec. 1974.

[27] P. Faurre, "Stochastic realization algorithms" in System Identification : advances and case studies, Mehra and Lainiotis Eds, Academic Press, 1976, ch. 1.

[28] B.L. Ho and R.E. Kalman, "Markov parameters, the moment problem, and minimal realizations", J. Control (SIAM), 4, 1965.

[29] D.G. Luenberger, "Introduction to dynamic systems: theory, models, and applications", Wiley, 1979.

[30] T. Kailath, "Linear Systems", Prentice-Hall, 1980.

[31] S.Y. Kung, K.S. Arun and D.V. Bhaskar Rao, "State space and singular value decomposition based approximation methods for the harmonic retrieval problem", Journal of Optical Society of America, Vol. 73, n° 12, Dec. 1983, pp. 1789-1811.

[32] J.P. Le Cadre and P. Ravazzola, "Utilisation de modélisation d'état en traitement d'antenne", XI Colloque GRETSI, Nice, 1987, pp. 385-388.

[33] K.S. Arun, D.V. Bhaskar Rao and S.Y. Kung, "A new predictive efficiency criterion for approximate stochastic realization", Proc. 22nd IEEE Conf. Decision and Control, San Antonio, TX, 1983, pp. 1353-1355.

[34] K.S. Arun, "A principal components approach to approximate modeling and ARMA Spectral estimation", Ph. D. Dissertation, University of Southern California, April 1984.

[35] U.B. Desai and D. Pal, "A realization approach to stochastic model reduction and balanced stochastic realization", Proc. 16th annual Conf. on Inf. Sciences and Systems, Princeton University, N.J. March 1982, pp. 613-620.

[36] A. Spataru, "Fondements de la théorie de la transmission de l'information", Compléments au traité d'électricité, Presses polytechniques romandes, 1987.

[37] H. Hotteling, "Relations between two sets of variables", Biometrika, Vol. 28, 1936, pp. 321-377.

[38] H. Akaike, "Markovian representation of stochastic processes by canonical variable", SIAM J. Control, Vol. 13, n° 1, January 1975, pp. 162-173.

[39] V.M. Adamjan, D.Z. Avov and M.G. Kvein, "Analytic properties of Schmidt pairs for a Hankel operator and the generalized Schur-Takagi problem", Math USSR Sbornik, Vol. 15, 1971, n° 1, pp. 31-73.

[40] N. Young, "An introduction to Hilbert spaces", Cambridge University Press, 1988.

[41] L.M. Silverman and M. Bettayeb, "Optimal approximation of linear systems", Proc. 1980 Joint Autom. Control Conf., paper FA8A.

[42] J.P. Le Cadre and P. Ravazzola, "Réduction de modèle et traitement d'antenne", XII Colloque GRETSI, Juan-les-Pins, 1989, Vol. 1 pp. 485-488.

[43] A. Björck and G.H. Golub, "Numerical methods for computing angles between linear subspaces", Math. Computation, Vol. 27, n° 123, July 1973, pp. 579-594.

CHAPTER 3

LOCALIZATION OF BROAD-BAND SOURCES WITH AN ARRAY OF UNKNOWN GEOMETRY

P. NICOLAS
SACLANT Underwater Research Center, La Spezia (Italie)

G. VEZZOSI *
Thomson-Sintra ASM, Valbonne (France)

*: now, with Université de Rennes (France)

LOCALISATION OF BROADBAND SOURCES WITH AN ARRAY OF UNKNOWN GEOMETRY

Philippe NICOLAS* and Georges VEZZOSI**

* SACLANT Undersea Research Centre, I-19026 La Spezia, Italy

** Thomson-Sintra ASM, Chemin des Travails, BP 53, 06801 Cagnes-sur-mer

1. Introduction

Array processing is a standard tool for source localisation in underwater acoustics, geophysics and radar. Spatial array processing consists of sampling the received wavefronts induced by radiating sources on a field of sensors, processing coherently the sensor outputs and producing an image of the source field.

In recent years, an extensive body of theory and practice has been developed for spatial array processing. The need for higher resolution of the source field image, has led to an evolution in these techniques, from classical beamforming to the CAPON estimator [1], autoregressive methods [2] and finally the high resolution methods based on the eigendecomposition of the cross-spectral matrix measured at the array output [3],[4],[5]. The latter methods take advantage of the knowledge of the noise field and lead to a resolution only limited by the integration time, when the noise field model matches the reality. This emphasises that the evolution of these techniques is closely associated with a better understanding of the sound propagation in the medium, and an improvement of the acoustical models.

It is usually assumed that the observed signal field is stationary and homogeneous, or quasi-stationary and quasi-homogeneous, over the observation time. This assumption excludes source and sensor motion over the integration time. Usually, the previous methods have assumed a known array geometry. If $\vec{r}_1, \vec{r}_2, ...\vec{r}_N$ are the positions of the N sensors, the steering vector associated with a monochromatic plane wave with the normalized wavevector $\vec{k}$ and the frequency f, in a non-dispersive medium, is given by

$$\underline{d}(\frac{2\pi f}{c}\vec{k};\vec{r}_2,...,\vec{r}_N) = \begin{pmatrix} 1 \\ \exp -i\frac{2\pi f}{c}\vec{k}\vec{r}_2 \\ \vdots \\ \exp -i\frac{2\pi f}{c}\vec{k}\vec{r}_N \end{pmatrix} = \begin{pmatrix} 1 \\ \exp -i2\pi f D_2 \\ \vdots \\ \exp -i2\pi f D_N \end{pmatrix}, \qquad (1.1)$$

where D_n is the propagation time delay of the plane wave between the nth sensor and the first sensor, taken as the reference ($\vec{r}_1 = 0$) and c is the sound velocity. This model also assumes that there is no differential attenuation of the plane wave between sensors. If the positions of the sensors are known, this vector depends only on one free parameter, $\vec{k}$.

The tendency to use arrays of numerous sensors limits the applicability of the present methods. In fact, "large array" is synonymous with uncertainty on the sensor positions, due to the array motion. Another practical situation where the previous methods cannot be applied is when the array is composed of sensors randomly distributed. This type of arrays has the advantage of being easily deployable, for example, a field of sonobuoys can be easily deployed from an airplane or a ship, but the exact locations are unknown.

The goal of the present work is to extend the eigenvector methods to the case of an array of sensors with unknown positions (random field) or known with uncertainties (large arrays). In this case, the steering vector $\underline{d}(\vec{k}, \vec{r_2}, ..., \vec{r_N})$ becomes a function of N free parameters. The problem of estimating the source bearings and the sensor positions is solvable if the number of sources which illuminate the medium is large enough: we shall see further on that three sources are necessary in the two-dimensional case and six sources are necessary in the three-dimensional case. The transmitted signals are assumed to be broadband processes; further, it is assumed that the signals received on the array are scaled versions of the signals transmitted by the sources, without any differential attenuation between two sensors. This last assumption restricts the size of the array. The signals and the background noise are uncorrelated. The noise itself is assumed to be spatially white, or its covariance can be estimated. Unlike the existing eigenvector techniques, the proposed method first estimates the differential time delays and is divided in two parts: multifrequency analysis and localisation.

Multifrequency analysis : This technique is a spectral multistep procedure and is based on the model of a nondispersive propagation medium described above. At any frequency f, the array response to the pth coherent source, $1 \leq p \leq P$, characterized by its set of time delays $D_{2p}, \ldots, D_{Np}$, is colinear to the steering vector

$$\underline{d}_p = [1, e^{i\theta_{2p}}, ..., e^{i\theta_{Np}}]^{\mathrm{T}}, \tag{1.2}$$

where

$$\theta_{np} = 2\pi f D_{np} \tag{1.3}$$

denotes the phase angle between the first and the nth sensors for the pth source, and

$$\underline{\theta}_p = [\theta_{2p}, ...\theta_{Np}]^{\mathrm{T}} \tag{1.4}$$

is the corresponding phase vector. The cross-spectral matrix of the observations is given by

$$\Gamma(f) = G(f)S(f)G^*(f) + a(f)I, \tag{1.5}$$

where $S(f)$ is the source spectral density matrix, $G(f)$ is the $N \times P$ transfer matrix of the source-array system whose pth column is the steering vector $\underline{d}_p(f)$. The quantity $a(f)$ is the power of the background noise assumed to be spatially white and $*$ denotes the transpose conjugate. It is also assumed that the sources are not coherent. For uncorrelated sources, the cross-spectral matrix reduces to

$$\Gamma(f) = \sum_{p=1}^{P} \gamma_p(f)\underline{d}_p(f)\underline{d}_p(f)^* + a(f)I \tag{1.6}$$

where $\gamma_p(f)$ is the power of the pth source. The first part of the multifrequency analysis consists of estimating the principal value of each phase angle from the cross-spectral matrix at each frequency. This procedure, which is called the monofrequency or narrow-band analysis, is described in the section 2.1. Taking advantage of the broadband property of the transmitted signals, and using the linearity of the absolute phase angles with frequency for constant time delay, one can link the phase vectors produced by the same sources. This procedure is described in section 2.2. The third part of the multifrequency analysis is the estimation of the time delays by a least square method and is developed in section 2.3.

Localisation: Let us assume that the sensors and the sources span a geometrical space of dimension d. Let $\vec{r}_1 = 0, \vec{r}_2, ..., \vec{r}_N$ be the sensor positions, where the first sensor is taken as a reference. The P broadband stationary plane waves radiated by the sources are defined by their direction vectors $\vec{k}_1, \vec{k}_2, ..., \vec{k}_P$. The time delays are then given by

$$\vec{r}_n.\vec{k}_p = cD_{np}, \quad 2 \leq n \leq N, \quad 1 \leq p \leq P. \tag{1.7}$$

Equation (1.7) can be rewritten in matrix form as follow

$$R^{\mathrm{T}}K = D, \tag{1.8}$$

where $D = [D_{np}]$ is the $(N-1) \times P$ matrix of time delays, R the $d \times (N-1)$ matrix of positions, and K is the $d \times P$ matrix of directions. Given an estimate $\hat{D}$ of D we can solve the non linear system (1.8) up to an orthogonal transformation. Therefore, without any other a-priori information, only the relative positions and directions can be determined. The estimation of the absolute directions and positions can be achieved if we know a-priori the positions of $d+1$ sensors. This is developed in section 3.

The algorithm has been tested on simulation and with at-sea experimental data. The results are shown in section 4.

The statistical performance of the algorithm are analysed both analytically and by Montecarlo simulations. The asymptotic variance of the time delay estimate is derived analytically in the single source case, and for multiple orthogonal sources. It is shown that the variance of the estimate reaches the Cramer- Rao lower bound. Montecarlo simulations confirm this result at high signal- to-noise ratio. At low signal-to-noise ratio, the estimates are subject to the threshold phenomenon due to the probability of large errors [6]. To quantify the statistical performance of the algorithm at low signal-to-noise ratio, we wish to compare the mean square error of the estimates with an appropriate lower SNR bound. Some of these bounds are available in the literature, among them the Barankin bound

[7] and the Ziv-Zakai lower bound [8]. The Ziv-Zakai lower bound has been applied with success to time delay estimation [9], provides the fundamental limitation in time delay estimation for a single parameter and describes the behaviour of the estimate around the SNR threshold. Therefore, the mean square error of the time delay estimate will be compared to the Ziv-Zakai lower bound for the single source and two sensors case. This is developed in section 5.

2. Multifrequency Analysis

The monofrequency analysis is not sufficient to estimate the differential time delays D_{np}. It only allows us to estimate the differential phase angles θ_{np}, between $-\pi$ and π, at a single frequency. Since

$$\theta_{np} = 2\pi f D_{np} \qquad \text{modulo} \qquad 2\pi,$$

the knowledge of θ_{np} at a single frequency defines D_{np} modulo $1/f$.

The goal of the multifrequency analysis is both to solve the previous ambiguity and to improve the time delay estimates by integrating the information on all the frequency bandwith of the source. Its implementation assumes

1) The estimation of the phase vectors at each single frequency, which is discussed in section 2.1;

2) The linkage of the phase vectors obtained at the different frequencies, which is discussed in section 2.2;

3) Once the links between phase vectors have been completed, the transformation of the differential phase angles into differential time delays which is discussed in section 2.3.

2.1 MONOFREQUENCY ANALYSIS: ESTIMATION OF THE PHASE VECTORS AT EACH SINGLE FREQUENCY

2.1.1 *Principle*

The method belongs to the class of the so-called signal subspace high resolution methods which process the observed cross-spectral matrix at the output of the N sensors. The principle of these methods is based on the separation of the observation space in two subspaces, the signal subspace and the noise subspace. In contrast to classical beamforming or adaptive array processing, this approach requires a more accurate modelisation of the medium and some parametric a-priori knowledge on the spatial coherence of the background noise. As an advantage, the so-called high resolution methods, asymptotically, have an infinite resolution. We first give an outline of the method; more details of each step will be presented in the following sections.

The estimated signal subspace is spanned by the eigenvectors associated with the P largest eigenvalues of the observation cross-spectral matrix $\hat{\Gamma}$. Let $(\underline{u}_1, \underline{u}_2, ..., \underline{u}_P)$ be these P orthonormalised eigenvectors stored in the $N \times P$ matrix U. We define the projection matrix onto the signal subspace by

$$\Pi = UU^*, \tag{2.1}$$

with $*$ denoting the conjugate transpose. Let $\underline{d}_{\underline{\theta}}$ be the steering vector model. For one of the P sources, this vector has the form (the frequency is omitted for simplicity)

$$\underline{d}_{\underline{\theta}} = \begin{pmatrix} 1 \\ \exp -i\theta_2 \\ \vdots \\ \exp -i\theta_N \end{pmatrix}, \tag{2.2}$$

where the differential phase angles are considered as free parameters, and $\underline{\theta}$ is the phase vector $[\theta_2, \ldots, \theta_N]^{\mathrm{T}}$. Note that there is no differential attenuation between sensors. We wish to estimate P such steering vectors.

The method for estimating the parameters θ_j, $2 \le j \le N$, consists of minimising the angle between the model vector and the estimated signal subspace. Since the norm of the vector $\underline{d}_{\underline{\theta}}$ is independent of $\underline{\theta}$, this minimisation is equivalent to maximise the squared norm of the projection of $\underline{d}_{\underline{\theta}}$ onto the signal subspace. The squared norm, $F(\theta)$, is defined by the function of $N-1$ variables

$$F(\underline{\theta}) = F(\theta_2, \ldots \theta_N) = \underline{d}_{\underline{\theta}}^* \Pi \underline{d}_{\underline{\theta}}. \tag{2.3}$$

Let $\underline{z} = (z_1, \ldots, z_P)^{\mathrm{T}}$ be a complex column vector of order P, where $z_p = x_p + iy_p$ and $\underline{v}$ a vector of the estimated signal subspace defined by

$$\underline{v} = [v_n]_{1 \le n \le N} = z_1 \underline{u}_1 + \ldots + z_P \underline{u}_P = U\underline{z}. \tag{2.4}$$

Maximising (2.3) is then equivalent to solving, in the least square sense, the following nonlinear system

$$U\underline{z} = \underline{d}_{\underline{\theta}}. \tag{2.5}$$

Because of the large number of unknown parameters in $\underline{d}_{\underline{\theta}}$, the optimisation of $F(\underline{\theta})$ requires numerical methods. Since the system is nonlinear, a descent algorithm is used. The initialisation is given by a solution $\underline{z}_0$ of the system

$$| U\underline{z} |^2 = \mathbf{1}_{N,1} \quad , \tag{2.6}$$

with, in general, $\mathbf{1}_{n,p}$ denoting the $n \times p$ matrix with all elements equal to one.

The nth row of (2.6) provides the equation

$$f_n(\underline{z}) = f_n(z_1, \ldots, z_P) = | v_n |^2 - 1 = 0, \tag{2.7}$$

where

$$1 \le n \le N, \quad , z_p = x_p + iy_p, \quad , 1 \le p \le P.$$

The quadratic system (2.7), with respect to the real unknowns (x_p, y_p), is solved in the least square sense by minimising

$$\Phi(\underline{z}) = \frac{1}{2} \sum_{n=1}^{N} f_n^2(\underline{z}) \tag{2.8}$$

which is achieved by a gradient method. This leads to the starting solution $\underline{z}_o$.

The solution of the system of equations (2.5) requires an additional constraint on the number P of sources and the number N of sensors. If we assume that the signal subspace has been perfectly estimated, this system has P exact solutions. The lefthand side has P complex unknowns $(z_1, \ldots, z_P)$, i-e, $2P$ real unknowns $(x_1, y_1, \ldots, x_P, y_P)$. The righthand side has $N-1$ real unknowns which are the phase angles $\theta_2, \ldots, \theta_N$. By separating real and imaginary parts, the system (2.6) is equivalent to $2N$ real equations. In order to have a finite number of solutions, the condition $2N \ge 2P + N - 1$ must be fulfilled, i-e,

$$N \ge 2P - 1.$$

If $N = 2P - 1$, this system presents many spurious solutions.

We now detail the three parts of the procedure: the estimation of the signal subspace, the solution of the quadratic system (2.6), and, the optimisation of the projection (2.5).

2.1.2 *Estimation of the signal subspace*

The cross-spectral matrix is estimated via a classical Fourier method. The Fourier transform of the observation $\underline{x}(t), -T/2 \le t \le T/2$ and the periodogram are computed as follows

$$\underline{x}(f) = \frac{1}{\sqrt{T}} \int_{-T/2}^{T/2} \underline{x}(t) \exp(-2i\pi f t) dt \tag{2.9}$$

$$I(f) = \underline{x}(f)\underline{x}(f)^*. \tag{2.10}$$

In order to reduce the bias of the Fourier transform, the observation can be tapered with a window (Hamming, Hanning,...). Finally, the cross-spectral matrix, at the given frequency, is computed by averaging the periodograms. It is well known that the density of the estimated cross-spectral matrix $\hat{\Gamma}$ can be approximated by the complex Wishart density with $K = WT$ degrees of freedom, where W is the spectral resolution of the Fourier transform. Assuming that K is an integer, the distribution of $\hat{\Gamma}$ is the same as the distribution of

$$\frac{1}{K} \sum_{k=1}^{K} \underline{x}_k \underline{x}_k^*, \tag{2.11}$$

where $(\underline{x}_k)_{1\leq k\leq K}$ denotes a sequence of K independent complex centered Gaussian vectors such that $\mathcal{E}\{\underline{x}_k\underline{x}_k^*\} = \Gamma$, $\mathcal{E}\{\underline{x}_k\underline{x}_k^T\} = 0$, where $\mathcal{E}$ stands for mathematical expectation, and T denotes the transpose.

When the number P of sources is unknown, the estimation of the signal subspace is equivalent to estimating the rank of the cross-spectral matrix of the signals. Some methods, based on the application of the information criteria of Akaike (AIC) and its Rissanen version (MDL) for model selection, have been developed, [10],[11]. They assume that the cross-spectral matrix of the observation, for independent signal and noise, is given by

$$\Gamma = \Gamma_S + aI, \tag{2.12}$$

where Γ_S is the cross-spectral matrix of the signals and aI is the cross-spectral matrix of the noise. For non coherent sources, the rank of Γ_S is equal to P, and the signal subspace is spanned by the eigenvectors associated with the P largest eigenvalues of Γ. Akaike proposed to select the rank P which yields the minimum of the function

$$AIC(P) = -\log p(\underline{x}, \hat{a}, \hat{\Gamma}_S) + m, \tag{2.13}$$

where $\hat{a}$ is the maximum likelihood estimate of the noise power, $\hat{\Gamma}_S$ is the maximum likelihood estimate of Γ_S and m is the number of free parameters in $\hat{\Gamma}_S$. $\underline{x}$ is the observation whose density function is given by

$$p(\underline{x}; a, \Gamma s) = \pi^{-NK}(det\Gamma)^{-K}\exp\{-\sum_{k=1}^{K}\underline{x}_k^*\Gamma^{-1}\underline{x}_k\}. \tag{2.14}$$

Rissanen proposed to select the rank P which yields the minimum of the function

$$MDL(P) = -\log p(\underline{x}, \hat{a}, \hat{\Gamma}_S) + \frac{1}{2}m\log K. \tag{2.15}$$

The maximum likelihood estimates $\hat{a}$ and $\hat{\Gamma}_S$ are respectively given by [11]

$$\hat{a} = \frac{1}{N-P}\sum_{j=P+1}^{N}\hat{\lambda}_j, \tag{2.16}$$

$$\hat{\Gamma}_S = \sum_{j=1}^{P}(\hat{\lambda}_j - \hat{a})\hat{\underline{u}}_j\hat{\underline{u}}_j^*, \tag{2.17}$$

where $\hat{\lambda}_1 \geq \hat{\lambda}_2 \geq \ldots\hat{\lambda}_N$ are the eigenvalues of the estimate $\hat{\Gamma}$ and $\hat{u}_j$ the associated eigenvectors. The log-likelihood function is given, up to a constant, by

$$\log p(\underline{x}, \hat{a}, \hat{\Gamma}_S) = -K(N-P)\log\hat{a} + K\sum_{j=P+1}^{N}\log\hat{\lambda}_j. \tag{2.18}$$

In both cases, the number of free parameters in the matrix Γ_S is equal to the number of free parameters in its Choleski factorisation. Since this factorisation is an $N \times P$ upper-triangular matrix with real diagonal coefficients and complex upper-diagonal coefficients, the number of free parameters is given by

$$m = P + 2(N-1) + \ldots + 2(N-P) = P(2N-P). \tag{2.19}$$

We exclude the parameters of the noise spectral matrix which are common to all orders. Then the criteria take the form

$$AIC(P) = (N-P)\log \hat{a} - \sum_{j=P+1}^{N} \log \hat{\lambda}_j + \frac{P(2N-P)}{K}, \tag{2.20}$$

$$MDL(P) = (N-P)\log \hat{a} - \sum_{j=P+1}^{N} \log \hat{\lambda}_j + \frac{P(2N-P)}{2K}\log K. \tag{2.21}$$

Wax and Kailath [11] showed that the AIC criterion yields an inconsistent estimate and tends to over-estimate the number of signals, while the MDL criterion provides a consistent estimate. In addition, some tests with experimental data confirm this result and therefore, we use the MDL criterion.

The assumption of spatially white noise does not constitute a drawback for these tests. If the noise is spatially correlated, and if a good estimate of its spectral matrix can be obtained, we come back to the previous case by prewhitening the observation matrix Γ. Given $\hat{\Gamma}_n$ an estimate of the cross-spectral matrix of the background noise, a prewhitening procedure can be the following:

1) Decompose $\hat{\Gamma}_n$ using the Choleski factorization

$$\hat{\Gamma}_n = L\Sigma L^*,$$

where Σ is a diagonal matrix and L is a lower triangular complex matrix.

2) Compute the matrix L_1 given by

$$L_1 = \sqrt{\Sigma}L^*$$

such that

$$\hat{\Gamma}_n = L_1^* L_1.$$

3) Compute the matrix

$$\begin{aligned}(L_1^*)^{-1}\hat{\Gamma}L_1^{-1} &= (L_1^*)^{-1}\hat{\Gamma}_s L_1^{-1} + (L_1^*)^{-1}\hat{\Gamma}_n L_1^{-1} \\ &= (L_1^*)^{-1}\hat{\Gamma}_s L_1^{-1} + I\end{aligned}$$

4) Apply the MDL criterion to the matrix $(L_1^*)^{-1}\hat{\Gamma}L_1^{-1}$.

Assuming that $\hat{\Gamma}_n$ has been estimated with an integration time much longer than the integration time used to estimate Γ, the MDL criterion defined by (2.21) can be applied without any modification. Otherwise, the statistics of $(L_1^*)^{-1}\hat{\Gamma}L_1^{-1}$ will be different from the statistics of $\hat{\Gamma}$ and the MDL criterion needs to be adjusted.

2.1.3 *Solution of the quadratic system*

The quadratic system (2.6) has been introduced to initialise the nonlinear system (2.5). Since there is no direct method for solving this quadratic system, we will use an iterative algorithm which requires itself a set of starting points. The computation of good starting points usually requires empirical methods specific to each problem. In our case the following procedure has turned out to provide satisfactory results.

Let us multiply the array U which contains the orhonormal basis of the signal subspace by $\sqrt(N)$, the norm of the steering vector model

$$\tilde{U} = \sqrt{N}U, \tag{2.22}$$

and consider the corresponding quadratic system

$$|\tilde{U}\underline{z}|^2 = \mathbf{1}_{N,1}, \tag{2.23}$$

where $\mathbf{1}_{n,p}$ denotes the $n \times p$ matrix with all elements equal to one. Assume now that the signal subspace has been identified perfectly and does not contains any spurious solution. Then the system (2.23) admits P exact distinct solutions $\underline{z}_1, \dots, \underline{z}_P$ which can be stored in a square matrix of order P

$$Z = [\underline{z}_1, \dots \underline{z}_P],$$

and (2.23) is rewritten in the equivalent form

$$|\tilde{U}Z|^2 = \mathbf{1}_{N,P}. \tag{2.24}$$

Since the steering vectors are assumed to be linearly independent, the matrix Z is invertible, and it is equivalent to compute P distinct solutions of (2.23) or an invertible matrix, solution of (2.24). According to (2.24), Z can be viewed as an operator which converts the array $\tilde{U}$ into an array with all elements equal to one in modulus. Following a current practice in numerical analysis, we try to gradually build up Z by means of plane transformations, which operate on column pairs of the array $\tilde{U}$, each transformation being optimised in order to make the columns on which it operates as close to the prescribed model as possible.

For analytical convenience, we retained the following transformations:

1) *Complex rotations:*

$$\begin{pmatrix} \cos\varphi & -\sin\varphi\exp\{-i\alpha\} \\ \sin\varphi\exp\{i\alpha\} & \cos\varphi \end{pmatrix}, \qquad |\varphi| \leq \frac{\pi}{4}. \tag{2.25a}$$

2) *Elementary transformations of the form:*

$$\begin{pmatrix} \cos\varphi & -i\epsilon\sin\varphi\exp\{-i\alpha\} \\ \sin\varphi\exp\{i\alpha\} & \cos\varphi \end{pmatrix}, \quad \epsilon = \pm 1, \quad |\varphi| \leq \frac{\pi}{4}. \tag{2.25b}$$

called the pseudorotations.

3) *Scale changes:* play a role of normalisation.

If the steering vectors were known to be orthogonal, transformations of the first kind would be sufficient for the initialisation. The need for transformations of the second and third kinds arises from the lack of orthogonality.

The initialisation step consists then in successively applying, to the columns of the array $\tilde{U}$, the following:

- a cycle of complex rotations which process the pairs of columns of $\tilde{U}$ in a given order, for instance, the order

$$(1,2),(1,3),\ldots,(1,P),(2,3)\ldots,(2,P),\ldots,(P-1,P); \tag{2.26}$$

- a cycle of pseudorotations which process the pairs of columns of $\tilde{U}$ in the same order; and

- a cycle of scale changes which treat the columns of $\tilde{U}$ in the natural order $1,\ldots,P$.

When the diagonal elements of the transformations of the first two kinds are close to zero, these transformations reduce nearly to a permutation of columns and a multiplication of each column by a phase factor. However, the inclusion of permutations in a sequence of transformations prescribed in a given order such as (2.26) ruins the sequence. For instance, if P equals 3, and if the processing of the pairs (1,2) and (2,3) of the original array is reduced merely to permutations, the pair (1,3) of the array will never be processed, while the pair (1,2) will be processed twice. To avoid this situation, the angle φ in (2.25) is constrained to be smaller than $\pi/4$ in modulus, as in the Jacobi method of diagonalisation.

The products of each elementary transformations are accumulated in the square complex matrix Z. The procedure is stopped when the total decrease of the criterion, defined as the Frobenius norm $|\tilde{U}|^2 - \mathbf{1}_{N,p}$, becomes, after each step, less than a preset threshold.

After convergence, the columns of $\sqrt{N}Z$ yields P starting points for the resolution of the quadratic system. The detailed derivations to obtain the formulas which optimise each elementary transformation are given in [12].

The solutions, in the least-square sense, of the quadratic system (2.7) are defined up to a phase factor: if $\underline{z} = (z_1, \ldots, z_P)^{\mathrm{T}}$ is a solution, the product $e^{i\beta}\underline{z}$ is also a solution. This fact causes the degeneracy of the matrices of partial derivatives near the solutions, and impairs drastically the convergence of descent algorithms. In order to avoid this situation, one force any nonnull component of the solution to be real. The best strategy would be to normalise the component of highest modulus. Since it is not possible to know the position of the component of highest modulus before computation, we normalise the component of highest modulus of the starting point. The normalisation is periodically reset during the iterations. By a suitable reordering of the variables, we can, without loss of generality, assume that the first component has been normalised. In this section, the symbol $\underline{z}$ denotes either of the following:

- the complex vector of order P whose first component is real

$$\underline{z} = [z_1, \ldots, z_P]^{\mathrm{T}}, \qquad \Im z_1 = 0; \tag{2.27a}$$

- the real vector of order $2P - 1$

$$\underline{z} = [x_1, x_2, y_2, \ldots, x_P, y_P]^{\mathrm{T}}. \tag{2.27b}$$

Furthermore, the quadratic system will be written briefly

$$\underline{f}(\underline{z}) = 0,$$

where $\underline{f} = [f_1, \ldots, f_N]^{\mathrm{T}}$, so that the criterion (2.8) becomes

$$\Phi(\underline{z}) = \frac{1}{2}\underline{f}^{\mathrm{T}}(\underline{z})\underline{f}(\underline{z}). \tag{2.28}$$

The quadratic system, successively initialised by the P starting vectors given by the initialisation step, is then solved by a classical descent algorithm [13]

$$\underline{z}^{(n+1)} = \underline{z}^{(n)} - \rho_n \underline{h}^{(n)},$$

where

$$\Phi(\underline{z}^{(n+1)}) \; < \; \Phi(\underline{z}^{(n)}).$$

Two search directions are used.

1) The beginning of the descent is performed in the Gauss-Newton direction. This means that the quadratic function $\underline{f}(\underline{z})$ is linearised

$$\underline{f}(\underline{z} - \underline{h}) \cong \underline{f}(\underline{z}) - \nabla \underline{f}\underline{h}, \tag{2.29}$$

where $\nabla \underline{f}$ is the matrix of partial derivatives of $\underline{f}$

$$\nabla \underline{f} = \begin{bmatrix} \frac{\partial f_1}{\partial x_1} & \frac{\partial f_1}{\partial x_2} & \frac{\partial f_1}{\partial y_2} & \cdots & \frac{\partial f_1}{\partial y_P} \\ \vdots & \vdots & \vdots & \ddots & \vdots \\ \frac{\partial f_N}{\partial x_1} & \frac{\partial f_N}{\partial x_2} & \frac{\partial f_N}{\partial y_2} & \cdots & \frac{\partial f_N}{\partial y_P} \end{bmatrix}.$$

The analytical expression of $\nabla \underline{f}$ is given in [12]. We solve, in the least-square sense, the over-determined linear system obtained by equating to zero the second member of (2.29)

$$\nabla \underline{f} \underline{h} = \underline{f}(\underline{z}).$$

2) The end of the descent is performed in the Hessian direction. This means that we expand the criterion $\Phi(\underline{z})$ up to the second order

$$\Phi(\underline{z} - \underline{h}) \cong \Phi(\underline{z}) - \underline{h}^{\mathrm{T}} \nabla \underline{f}^{\mathrm{T}} \underline{f}(\underline{z}) + \frac{1}{2} \underline{h}^{\mathrm{T}} H \underline{h}, \tag{2.30}$$

where H is the Hessian of the criterion Φ given by

$$H = \nabla \underline{f}^{\mathrm{T}} \nabla \underline{f} + \sum_{n=1}^{N} (\nabla^2 f_n) f_n.$$

$\nabla^2 f_n$ is the Hessian of f_n. We minimise the quadratic form of the second member of (2.30), which yields the linear system

$$H \underline{h} = \nabla \underline{f}^{\mathrm{T}} \underline{f}(\underline{z}).$$

Since the system of equations is quadratic, the function

$$g(\lambda) = \Phi(\underline{z} - \rho \underline{h})$$

is, for fixed $\underline{h}$, a polynomial of fourth degree in ρ. This allows us to determine the optimum stepsize ρ within a finite number of steps once the search direction $\underline{h}$ has been computed.

2.1.4 *Optimisation of the projection*

This section sets up the criterion optimising the projection, gives asymptotic results on the angle between the true steering vector and the estimated signal subspace, and the error covariance matrix of the phase vector estimates.

a) *The Criterion which optimises the projection:*

Let $\Pi = UU^*$ be the projection matrix onto the estimated signal subspace, and let $\underline{v}$ be the corresponding projection of the steering vector model (2.2). We have

$$\begin{aligned}\underline{v} = [v_n]_{1\leq n\leq N} &= \Pi \underline{d}_{\underline{\theta}} = UU^* \underline{d}_{\underline{\theta}} \\ &= U\underline{z} = z_1\underline{u}_1 + \ldots + z_P\underline{u}_P,\end{aligned} \tag{2.31}$$

where the complex vector of order P

$$\underline{z} = [z_1, \ldots, z_P]^{\mathrm{T}} = U^* \underline{d}_{\underline{\theta}} \tag{2.32}$$

defines the component of the projection in the signal subspace. Maximising the projection is equivalent to maximising the squared norm

$$F(\underline{\theta}) = \|\underline{v}\|^2 = \|\underline{z}\|^2 = \underline{d}_{\underline{\theta}}^* \Pi \underline{d}_{\underline{\theta}}. \tag{2.33}$$

Following some rather extensive derivations [12], it is shown that the maximisation of (2.33) is equivalent to minimising the criterion

$$\Psi(\underline{z}) = \frac{1}{2}\sum_{n=1}^{N}(\mid v_n \mid -1)^2. \tag{2.34}$$

The form (2.34) of the criterion shows that the optimisation of the projection is equivalent to the solution in the least-square sense of

$$\mid v_n \mid = 1, \qquad 1 \leq n \leq N,$$

while the quadratic system was obtained by solving

$$\mid v_n \mid^2 = 1, \qquad 1 \leq n \leq N.$$

The ambiguity due to the phase factor can be reduced as in the quadratic system, by forcing the component of highest modulus of the starting point to be real. Then the system is solved by Newton's method, using the solution of the quadratic system as starting points.

The quadratic system (Eq 2.6) and the system optimising the projection (Eq 2.5) have an exact solution if there exists a point $\underline{z}$ such that the corresponding $\underline{v} = U\underline{z}$ satisfies $\mid v_n \mid = 1$ for all n. It is obvious that both systems have the same set of exact solutions. In particular, when the signal subspace is identified perfectly, the systems (2.5) and (2.6) are equivalent. This equivalence is not maintained when the estimation of the signal subspace becomes noisy, and in this case the solution in the least-square sense of the quadratic system does not optimise the projection.

b) *Angle between the true steering vector and the estimated signal subspace*

When a solution of the system which optimises the projection has been found, one must check if this solution corresponds to a source or is a spurious solution. A natural way to do so, is to compare the angle between the associated steering vector and the estimated signal subspace to a threshold, depending upon the accuracy of the estimated cross-spectral matrix. This angle $\alpha(\underline{\theta})$ is defined by its sine as follow

$$\begin{aligned}\sin^2\alpha(\underline{\theta}) &= 1 - \underline{d}_{\underline{\theta}}^* \Pi \underline{d}_{\underline{\theta}}/N \\ &= 1 - \underline{z}^* U_0^* \Pi U_0 \underline{z}/N,\end{aligned} \tag{2.34}$$

where U_0 denotes the array which stores the orthogonal basis of the true signal subspace. Our purpose is to obtain the first term of the expansion of $\alpha(\underline{\theta})$. The derivation relies upon a perturbation analysis of Hermitian matrices. The density of the estimated spectral matrix $\hat{\Gamma} = \Gamma + \Delta\Gamma$ is approximated by a complex Wishart density with K degrees of freedom, so that [14]

$$\mathcal{E}\{\Delta\Gamma(j,k)\Delta\bar{\Gamma}(m,n)\} = \Gamma(j,m)\bar{\Gamma}(k,n)/K \qquad 1 \le j,k,m,n \le N, \tag{2.35}$$

where $^-$ denotes the conjugate.

Given the spectral decomposition of Γ

$$\Gamma = V\Lambda V^*,$$

where

$$\Lambda = diag(\lambda_1, \ldots, \lambda_N), \qquad \lambda_{P+1} = \ldots = \lambda_N = a,$$

let us define the normalised perturbation

$$\Delta E = \Lambda^{-1/2} V^* \Delta\Gamma V \Lambda^{-1/2}.$$

Replacing Γ by the identity matrix in (2.35), yields

$$\mathcal{E}\{\Delta E(j,k)\Delta\bar{E}(m,n)\} = \delta_{jm}\delta_{kn}/K, \tag{2.36}$$

where δ denotes the Kroenecker symbol. Following derivations based on the expansion of the P greatest eigenvectors and the projection matrix Π up to the second order [12], we show that the sine of the angle between the exact steering vector model and the estimated signal subspace is given by

$$sin^2\alpha(\underline{\theta}) = \frac{1}{N} \sum_{n=P+1}^{N} | Y_n |^2,$$

where

$$Y_n = \sum_{j=1}^{P} z_j \frac{\sqrt{\lambda_j a}}{\lambda_j - a} \Delta E(n,j), \qquad P+1 \le n \le N. \tag{2.37}$$

Under the Gaussian assumption for $\Delta E(n,j)$, each variable Y_n is a complex circular variable such that

$$\mathcal{E}\{|Y_n|^2\} = \frac{1}{K}\sum_{j=1}^{P} |z_j|^2 \frac{a\lambda_j}{(a-\lambda_j)^2}. \tag{2.38}$$

All the variables are independant and $\sin^2\alpha(\underline{\theta})$ is the sum of squares of $N-P$ variables of this type. Therefore, the random variable $\sin^2\alpha(\underline{\theta})$ has a $t\chi^2_\nu$ distribution, where

$$\nu = 2(N-P), \qquad t = \frac{1}{2NK}\sum_{j=1}^{P} |z_j|^2 \frac{a\lambda_j}{(a-\lambda_j)^2}. \tag{2.39}$$

This allows us to set up confidence intervals for testing the acceptability of any given solution of the system which optimises the solution.

c) *Error covariance matrix of the phase vectors*

Let $\hat{\underline{\theta}}$ be the estimate of the true phase angle vector $\underline{\theta}$. The derivation of the error covariance matrix of the estimate $\hat{\underline{\theta}}$ is performed using the first order approximation $\hat{\underline{\theta}} = \underline{\theta} + \Delta\underline{\theta}$. The vector $\hat{\underline{\theta}}$ annihilates the gradient $\nabla\underline{F}$ of the squared norm of the projection defined in (2.33)

$$(\nabla\underline{F})(\hat{\underline{\theta}}) = \underline{0}.$$

Linearising the gradient around the true value $\underline{\theta}$ yields

$$(\nabla\underline{F})(\underline{\theta}) + (\nabla^2 F)(\underline{\theta})\Delta\underline{\theta} = 0.$$

Assuming that the inverse of $\mathcal{A}$ exists, the error covariance matrix is given by

$$\mathcal{E}\{\Delta\underline{\theta}\Delta\underline{\theta}^{\mathrm{T}}\} = \mathcal{A}^{-1}\mathcal{B}\mathcal{A}^{-1},$$

where $\mathcal{A}$ and $\mathcal{B}$ are the square matrices of order $N-1$ given by

$$\mathcal{A} = [a_{jk}] = -\left[\mathcal{E}\{\frac{\partial^2 F}{\partial\theta_j\partial\theta_k}\}\right]_{2\le j,k\le N},$$

$$\mathcal{B} = [b_{jk}] = \left[\mathcal{E}\{\frac{\partial F}{\partial\theta_j}(\underline{\theta})\frac{\partial F}{\partial\theta_k}(\underline{\theta})\}\right]_{2\le j,k\le N}.$$

The derivations of the partial derivatives show that $\mathcal{A}$ and $\mathcal{B}$ are respectively defined by [12]

$$a_{jk} = 2\Re_e\{e^{-i(\theta_j-\theta_k)}e_j^{\mathrm{T}}\Pi_0^{\perp}e_k\},$$
$$b_{jk} = \mathcal{E}\{|Y_n|^2\}a_{jk},$$

where e_j stands for the j-unit coordinate vector and $\Pi_0^\perp = I - \Pi_0$ is the orthogonal projection matrix to $\Pi_0 = U_0 U_0^*$, Π_0 being the exact signal subspace projection matrix. The quantity Y_n is given by (2.37). The expressions of a_{jk} and b_{jk} show that the matrices $\mathcal{A}$ and $\mathcal{B}$ are colinear and this gives the asymptotic error covariance matrix at the phase angle estimates

$$\mathcal{E}\{\Delta\underline{\theta}\Delta\underline{\theta}^{\mathrm{T}}\} = \frac{1}{K}\sum_{j=1}^{P} | z_j |^2 \frac{a\lambda_j}{(a-\lambda_j)^2}\mathcal{A}^{-1}. \tag{2.40}$$

In the case of orthogonal sources, the result assumes the form

$$\mathcal{E}\{\Delta\underline{\theta}\Delta\underline{\theta}^{\mathrm{T}}\} = \frac{1}{2K}\sum_{p=1}^{P} | z_p |^2 \frac{a\lambda_p}{(a-\lambda_p)^2}(I + \mathbf{1}_{N-1,1}\mathbf{1}_{N-1,1}^{\mathrm{T}}) = C. \tag{2.41}$$

2.2 LINKAGE OF THE PHASE VECTORS

2.2.1 *Principle*

The multifrequency analysis consists of repeating the monofrequency processing on a set of equispaced frequency bins with frequency spacing Δf

$$\{f_m = m\Delta f, \quad 1 \le m \le M\}$$

the choice of Δf depends on the time delays involved in the field of sensors. Let $D_{\max}$ be the maximum time delay which can be observed; $D_{\max}$ is determined by the size of the sensor field and the speed of sound, i.e. by the distance between the further sensor and the reference one. The condition

$$\Delta f . D_{\max} < \frac{1}{2} \tag{2.42}$$

is required to enable the estimation of the time delays from the phase angles. Violation of this condition implies:

1) the impossibility of recovering $D_{\max}$ without ambiguity, since the series

$$\{e^{2i\pi f_m D_{\max}}, \quad f_m = m\Delta f\}$$

gives $D_{\max}$ up to $1/\Delta f$.

2) a large bias on the nondiagonal elements of the estimated cross-spectral matrix which contain the complex exponential

$$e^{i2\pi f_m D_{\max}};$$

The cross-spectral matrix estimates $\hat{\Gamma}(f_m)$ are assumed to be statistically independent (this condition is asymptotically satisfied for large WT, when $\Gamma(f_m)$ is estimated by a

Fourier method). The monofrequency processing of the matrix $\hat{\Gamma}(f_m)$ provides $Q(f_m)$ phase vectors which include the P true phase vectors and $Q(f_m) - P$ spurious phase vectors. We recall that the spurious vectors can occur from a too weak overdetermination of the system (2.5), i.e., a number of sensors too small compared to the number of sources. Figure 1 illustrates the multifrequency analysis and the chaining of the phase vectors.

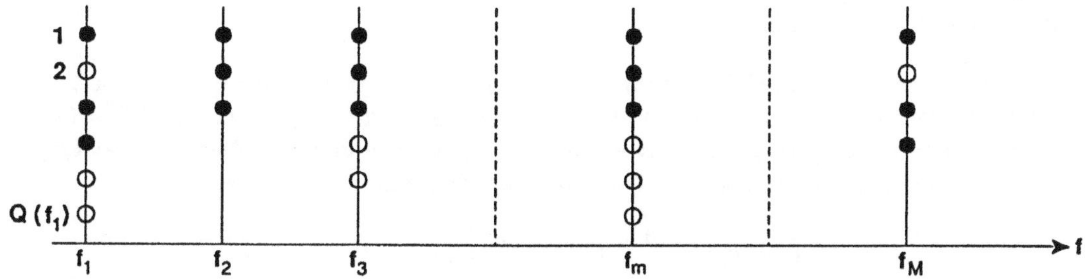

Fig. 1

The x-axis represents frequency and the y-axis represents the phase vectors estimated at each frequency. The solid and the open dots denote respectively the true and the spurious phase vectors. The phase vectors are ordered by increasing value of the sine of the angles between the corresponding steering vectors and the estimated signal subspace. The chaining procedure consists of connecting the phase vectors which correspond to the same source.

The estimates of the cross-spectral matrix, the signal subspace and the vectors of complex exponential $\underline{d}_\theta$ close to this subspace, are all continuous functions of frequency. Therefore, the chaining of the phase vectors found at the different frequencies is typically a continuity problem, and the chaining is always possible if the frequency analysis increment Δf is small enough. The suppression of the spurious vectors and the ambiguities coming from the crossing of the phase vectors can be easily resolved by using the linearity of the phase angles with frequency. In the following, a phase vector is identified with a pair of integers (a, m), where m is the frequency bin and a is the number of the phase vector in the set of $Q(f_m)$ phase vectors found at the frequency f_m. A chain of phase vectors is then a sequence of integers

$$(a_1, a_2, 0, \ldots, a_m, \ldots, a_M), \tag{2.43}$$

where the zero means that the chain contains no element at the corresponding frequency.

2.2.2 *The Elementary Triplets [15]*

Let $\phi(\theta)$ be the usual sawtooth function

$$\begin{cases} \phi(\theta) = \theta \quad \text{for} -\pi < \theta < \pi, \\ \phi(\theta + 2\pi) = \phi(\theta), \end{cases}$$

which converts any phase angle θ into its principal value. If $\underline{\theta} = [\theta_n]_{2 \leq n \leq N}$ is a vector, we note $\phi(\underline{\theta})$ the vector $[\phi(\theta_n)]_{2 \leq n \leq N}$.

Consider three equispaced frequency bins $f < f' < f''$ with frequency spacing $q\Delta f$, where q is an integer called the order of the triplet. Let $Q = Q(f)$, $Q = Q'(f')$, $Q'' = Q(f'')$ be the complete number of phase vectors found at these frequencies. Assume that $QQ'Q''$ is different from zero, and choose at each of these frequencies the phase vectors

$$\hat{\underline{\theta}} = [\hat{\theta}_n], \quad \hat{\underline{\theta}}' = [\hat{\theta}'_n], \quad \hat{\underline{\theta}}'' = [\hat{\theta}''_n], \quad 2 \leq n \leq N,$$

which are respectively indexed by a, b, c. Assuming that these vectors are not spurious solutions, they can be written up to 2π as the sum of the absolute phase angles and an error

$$\hat{\underline{\theta}} = \phi(\underline{\theta} + \Delta\underline{\theta}), \quad \mathcal{E}\{\Delta\underline{\theta}\Delta\underline{\theta}^{\mathrm{T}}\} = C,$$

with similar relations at frequencies f' and f''. Consider the vector

$$\underline{\Delta} = \phi(\hat{\underline{\theta}} + \hat{\underline{\theta}}'' - 2\hat{\underline{\theta}}') = \phi(\underline{\theta} + \underline{\theta}'' - 2\underline{\theta}' + \Delta\underline{\theta} + \Delta\underline{\theta}'' - 2\Delta\underline{\theta}'), \tag{2.44}$$

and neglect the case where $\underline{\theta} + \underline{\theta}'' - 2\underline{\theta}'$ has at least one of its components close to $(2k+1)\pi$. Then $\underline{\Delta}$ is asymptotically Gaussian, with mean $\phi(\underline{\theta} + \underline{\theta}'' - 2\underline{\theta}')$ and since the cross-spectral matrices estimated at f, f', f'' are statistically independent, its covariance is given by

$$\Omega = \Sigma + \Sigma' + 4\Sigma''. \tag{2.45}$$

As the absolute phase angles vary linearly with the frequency, the mean of $\underline{\Delta}$ equals zero when the three vectors $(\hat{\underline{\theta}}, \hat{\underline{\theta}}', \hat{\underline{\theta}}'')$ come from the same source. Thus, finding good triplets amounts to testing the nullity of the mean of a Gaussian vector with a known covariance. The optimal test (UMP invariant) is based on the comparison to a threshold of the statistics

$$\mu(a, b, c) = \underline{\Delta}^{\mathrm{T}} \Omega^{-1} \underline{\Delta}, \tag{2.46}$$

which under the null hypothesis has a central χ^2 distribution with $N-1$ degrees of freedom. This suggests a possible solution: for each of the $QQ'Q''$ triplets (a, b, c) which can be constructed with the phase vectors found at f, f' and f'', compute $\underline{\Delta}$ by means of (2.44), Ω by means of (2.45), where the unknown quantities are replaced by their estimates. Then compute the statistics $\mu(a, b, c,)$ by means of (2.46) and compare it with a χ^2_{N-1}.

At low signal to noise ratio, the conditions of asymptotic normality of the phase angles are not always fulfilled, and the simulations show that some triplets, although correct, are incompatible with a χ^2_{N-1}. Therefore, the procedure is modified as follows. Let (a_0, b_0, c_0) be a correct triplet. Assume that none of the points a_0, b_0, c_0 are common to more than one source. Then, even if the value of $\mu(a_0, b_0, c_0)$ is inconsistent with a χ^2_{N-1}, it can be inferred that the path (a_0, b_0, c_0) minimises $\mu(a, b, c,)$ among all the paths between f and f'' going respectively through a_0, b_0 and c_0. This suggests retaining only the triplets which satisfy the condition

$$\mu(a_0, b_0, c_0) = \inf_{b,c} \mu(a_0, b, c) = \inf_{a,c} \mu(a, b_0, c) = \inf_{a,b} \mu(a, b, c_0) \tag{2.47}$$

The computation of $\mu(a, b, c)$ at all frequencies, all orders and for all triplets is numerically too expensive. In practice, the computational complexity is significantly reduced if the complete whitening based on the inverse Ω^{-1} is replaced by an approximate one, defined by the matrix

$$B = (I + \mathbf{1}_{N-1,1}\mathbf{1}^{\mathrm{T}}_{N-1,1})^{-1} = I - \mathbf{1}_{N-1,1}\mathbf{1}^{\mathrm{T}}_{N-1,1}/N, \tag{2.48}$$

derived for the case of a single source (see 2.41). One then computes

$$\underline{\Delta}^{\mathrm{T}} B \underline{\Delta} = \sum_{n=2}^{N} \Delta_n^2 - (\sum_{n=2}^{N} \Delta_n)^2/N,$$

$$\mathcal{E}\{\underline{\Delta}^{\mathrm{T}} B \underline{\Delta}\} = \text{trace } (\Omega B)$$

and replaces everywhere μ by $\mu' = \underline{\Delta}^{\mathrm{T}} B \underline{\Delta} / \mathcal{E}\{\underline{\Delta}^{\mathrm{T}} B \underline{\Delta}\}$.

2.2.3 *Chaining the Phase Vectors*

At low signal to noise ratio, an error percentage of a thirty percent (i.e. linking triplets which do not correspond to any source) can be reached, and therefore a technique based only the fusion of triplets of order one at successive frequency would fail. As an alternative, the chaining procedure performs the fusion of triplets of different orders q, which leads to the so-called elementary chains of the form (2.43). These elementary chains are tested to ensure that they correspond to continuity of phase angles associated with a source. Then, each validated elementary chain is continued by filling the holes (0 in the chain) at frequencies where no-phase vector has yet been assigned. The chaining algorithm is an iterative continuation of the elementary chains and is divided in three steps: the initialisation or validation of the elementary triplets, the fusion of the validated triplets and the continuation of the elementary chains. The three parts of the algorithm are now detailed.

a) Initialisation: validation of the triplets

Any triplet of order q (a, b, c) defines two links of order $q, (a, b)$ and (b, c), and one link of order $2q, (a, c)$. The triplet (a, b, c) is said to be validated if the link (a, c) and one of the two links (a, b) and (b, c) are found in other triplets, and if (a, b, c) is compatible with all triplets going through a, b or c. Figure 2 shows an example of a triplet validation.

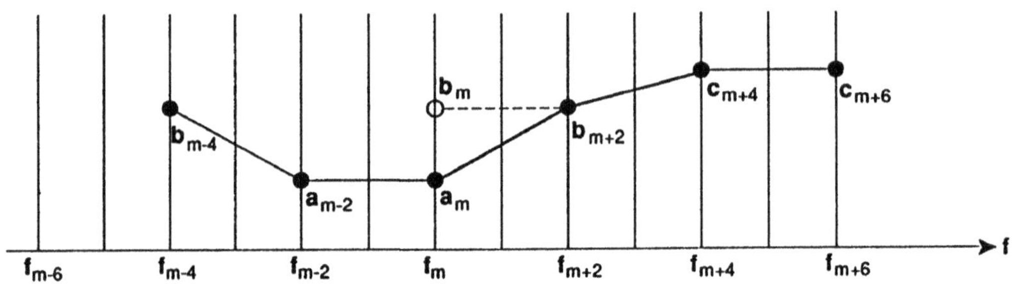

Fig. 2

Let us consider the triplet (a_m, b_{m+2}, c_{m+4}) where a_m is the ath phase vector found at the frequency f_m, b_{m+2} is the bth phase vector found at the frequency f_{m+2} and c_{m+4} is the cth phase vector found at the frequency f_{m+4}. The triplet (a_m, b_{m+2}, c_{m+4}) is of order 2 and we assume that it corresponds to a source, i.e. a_m, b_{m+2} and c_{m+4} are true phase vectors. Let b_{m-4} be the bth phase vector found at the frequency f_{m-4} associated with the same source, such that the link (a_m, c_{m+4}) also belongs to the triplet (b_{m-4}, a_m, c_{m+4}) of order 4. Then the first condition of the validation is fulfilled. Now assume that the ath phase vector found at f_{m-2} and the cth phase vector found at the frequency f_{m+6} also corresponds to the same source. Then the links (a_m, b_{m+2}) and (b_{m+2}, c_{m+4}) are respectively included in the triplets (a_{m-2}, a_m, b_{m+2}) and $(b_{m+2}, c_{m+4}, c_{m+6})$ of order 2 and the second condition is fulfilled. Furthermore, (a_m, b_{m+2}, c_{m+4}) is compatible with the following triplets: $(a_{m-2}, a_m, b_{m+2}), (b_{m+2}, c_{m+4}, c_{m+6}), (a_{m-2}, b_{m+2}, c_{m+6}), (b_{m-4}, b_{m+2}, c_{m+6})$. Now if the bth phase vector estimated at the frequency f_m was part of the triplet (b_m, b_{m+2}, c_{m+4}), the triplet (a_m, b_{m+2}, c_{m+4}) could not be validated since it would be incompatible with (b_m, b_{m+2}, c_{m+4}): a_m and b_m are at the same frequency and different.

Therefore the first two conditions of the validation of a triplet ensure that this triplet is not isolated, but belongs to a longer chain of phase vectors. It eliminates triplets of spurious phase vectors and narrow-band sources present only on a few frequency bins. The third condition guarantees that the links $(a_m, b_{m+2}), (b_{m+2}, c_{m+4}), (a_m, c_{m+4})$ are not inconsistent with other links of the kind $(a_m, b), (a, b_{m+2}), (b_{m+2}, c), (a, c_{m+4})$ or (a_m, c). In the initialisation step, we look for all the validated triplets in the frequency band $[f_1, f_M]$.

b) Fusion of the validated triplets:

Whenever two chains (i.e. the validated triplets) are compatible and have at least two non vanishing elements in common, they are merged and treated as a single chain. The fusion procedure provides a set of chains called the elementary chains.

c) Continuity of the elementary chains:

Consider a chain such that, for some frequency f, no phase vector has yet been assigned, as represented in Figure 3.

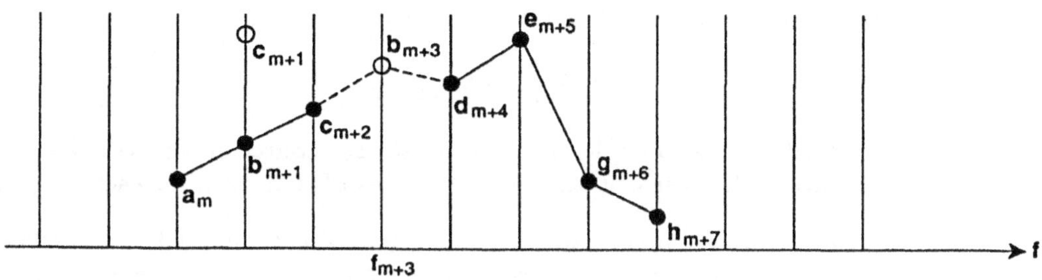

Fig. 3

The chain $(a_m, b_{m+1}, c_{m+2}, 0, d_{m+4}, e_{m+5}, g_{m+6}, h_{m+7})$ contains a zero at the frequency f_{m+3}. In order to fill this hole, we search, at the frequency f_{m+3}, the phase vectors which are connected to the chain by means of triplets. For instance, assume that the phase vector b_{m+3} belongs to the triplets $(a_m, b_{m+3}, g_{m+6}), (b_{m+1}, b_{m+3}, e_{m+5}), (b_{m+3}, e_{m+5}, h_{m+7})$. Since b_{m+3} is related three times to the incomplete chain, we say that it is a good candidate for the missing phase vector at the frequency f_{m+3}. This suggests an empirical method to continue the incomplete chains: the phase vector estimated at the frequency f_{m+3} are classified in decreasing order of the number of their links with the chain. If the first vector in the classification presents a number of links much greater than the second one, this vector is considered as a possible new element of the chain. However, before including definitively the vector in the chain, we must verify that b_{m+3} is not a crossing point. A crossing point is a phase vector which is connected to other phase vectors of different sources. For instance, b_{m+3} would be a crossing point if it was also linked to c_{m+1} (see Figure 3). The detection of crossing points is completed by looking at the pairs (c, b_{m+3}) and (b_{m+3}, c). If the chain made of these pairs is incompatible with $a_m, b_{m+1}, c_{m+2}, 0, d_{m+4}, e_{m+5}, g_{m+6}, h_{m+7})$, we decide that b_{m+3} is a crossing point and therefore is not included as a new element.

The new chains obtained after continuity are fusioned again using the procedure described in b) and the continuity is applied, and so on, until the chains become stable.

2.3.1 Principle

Assume that a chain of phase vectors associated with a source has been found which lie between the frequencies $f_1 = m_1\Delta f$ and $f_2 = m_2\Delta f$. Consider the nth component of this sequence ($2 \leq n \leq N$):

$$[\hat{\theta}_n(m_1), \ldots, \hat{\theta}_n(m_2)], \tag{2.49}$$

to which corresponds the sequence of absolute exact phase angles

$$[2\pi m \Delta f D_n]_{m_1 \leq m \leq m_2}, \tag{2.50}$$

where D_n is the differential time delay of the considered source on the nth sensor. If the chaining procedure performed well, there are P chains of such form for each sensor.

Given a sequence of phase angles, we want to estimate the time delay D_n. There are basically two methods for solving the problem. At high signal to noise ratio, the phase angles are estimated with a good precision, the chaining of the phase vectors is perfectly achieved and no phase angle is missing in the sequence (2.49). Therefore a fast procedure can be designed by unwrapping the phase angles and fitting a linear regression to the unwrapped phase angles [15]. This method estimates the complete time delay vector $\underline{D} = [D_2, \ldots, D_n]$ and therefore takes into account the correlation between the individual time delays D_n, $2 \leq n \leq N$. At low signal to noise ratio, some phase angles can be missing in the sequence (2.49) such that the phase difference between elements at two adjacent frequencies can be greater than π. In that case, the phase unwrapping becomes impossible. Also, even if the sequence is complete, the phase unwrapping suffers a considerable loss of efficiency due to the large errors on the phase angle estimates [16]. Therefore, in practice, it is preferable to estimate the time delays by minimising the least square criterion,

$$\Psi(D_n) = \sum_{m=m_1}^{m_2} | e^{i\hat{\theta}_n(m)} - e^{i2\pi m\Delta f D_n} |^2 / \text{var}\{\hat{\theta}_n(m)\}, \quad 2 \leq n \leq N, \tag{2.51}$$

which amounts to minimising

$$\psi(D_n) = \sum_{m=m_1}^{m_2} \cos(\hat{\theta}_n(m) - 2\pi m\Delta f D_n)/\text{var}\{\hat{\theta}_n(m)\}, \quad 2 \leq n \leq N. \tag{2.52}$$

In practice, the function $\psi(D_n)$ can be computed by FFT. In order to have a good precision on the time delay estimate, we zero-pad the sequence (2.49) with a sufficient number of zeros before performing the FFT. The variance of the phase angle var $\{\hat{\theta}_n(m)\}$ is obtained using (2.40) with the different parameters replaced by their estimates. The search for the maximum of the function $\psi(D)$ is achieved in the a-priori region $\left[-\frac{1}{2\Delta f}, \frac{1}{2\Delta f}\right]$, where

$1/2\Delta f$ corresponds to the maximum time delay which can be estimated. This second procedure estimates the time delays D_n, $2 \leq n \leq N$ separately and can be considered as suboptimal in comparison to the first one. However, it will be shown that the two procedures are asymptotically equivalent in the case of P orthogonal sources or in the single source case. They have the same aymptotic statistical performance and lead to the same variance of the time delay estimates.

Let us now derive the error covariance matrix of the time delay vector, estimated by the phase unwrapping method for the case of P orthogonal sources.

2.3.2 *Error covariance matrix of the time delay vector estimate*

Let

$$\underline{\hat{\theta}}(m) = [\hat{\theta}_2(m), \ldots, \hat{\theta}_N(m)]^{\mathrm{T}}, \quad m_1 \leq m \leq m_2 \tag{2.53}$$

be the phase vector after unwrapping (estimate of the absolute phase vector) at the frequency f_m, with its error covariance matrix $C(m)$. Note that the error covariance matrix of the absolute phase angles is equal to the error covariance matrix of their principal values. Since the asymptotic distribution of the phase vector estimate is Gaussian, the estimation of the time delay vector

$$\underline{D} = [D_2, \ldots, D_N]^{\mathrm{T}}$$

is performed by maximising the log-likelihood function, which leads to minimising the quadratic form

$$\frac{1}{2} \sum_{m=m_1}^{m_2} [\underline{\hat{\theta}}(m) - 2\pi m \Delta f \underline{D}]^{\mathrm{T}} C^{-1}(m) [\underline{\hat{\theta}}(m) - 2\pi m \Delta f \underline{D}].$$

The minimisation of this quadratic form is equivalent to solving the linear system

$$\mathcal{F}\underline{D} = \Theta, \tag{2.54}$$

where

$$\mathcal{F} = 2\pi\Delta f \sum_{m=m_1}^{m_2} m^2 C^{-1}(m),$$

$$\Theta = \sum_{m=m_1}^{m_2} m C^{-1}(m) \underline{\hat{\theta}}(m).$$

For orthogonal sources, the matrices $C(m)$, $m_1 \leq m \leq m_2$, are colinear and are given by (2.41)

$$C(m) = \beta(m) C_0$$

where

$$\beta(m) = \frac{1}{K} \sum_{p=1}^{P} \mid z_p(f_m) \mid^2 \frac{a(f_m)\lambda_p(f_m)}{(a(f_m) - \lambda_p(f_m))^2},$$

$$C_0 = \frac{1}{2}(I + \mathbf{1}_{N-1,1}\mathbf{1}_{N-1,1}^{\mathrm{T}}). \tag{2.55}$$

Thus, the linear system (2.54) reduces to a diagonal system whose solution is given by

$$\underline{\hat{D}} = \frac{1}{2\pi\Delta f \sum_{m=m_1}^{m_2} \frac{m^2}{\beta(m)}} \sum_{m=m_1}^{m_2} \frac{m}{\beta(m)} \underline{\hat{\theta}}(m). \tag{2.56}$$

The error covariance matrix of the time delay vector has the following form, where $\beta(m)$ has been replaced by its value given by (2.55),

$$\mathcal{E}\{\Delta\underline{D}\Delta\underline{D}^{\mathrm{T}}\} = \frac{1}{2K \sum_{m=m_1}^{m_2} (2\pi f_m)^2 \left(\sum_{p=1}^{P} | z_p(f_m) |^2 \frac{a(f_m)\lambda_p(f_m)}{\left(a(f_m) - \lambda_p(f_m)\right)^2} \right)^{-1}} (I + \mathbf{1}_{N-1,1}\mathbf{1}_{N-1,1}^{\mathrm{T}}). \tag{2.57}$$

Let us now consider the single source case. If $\gamma(f_m)$ is the power of the single source at the frequency f_m, the eigenvalue associated with the source is given by

$$\lambda(f_m) = a(f_m) + N\gamma(f_m) = a(f_m)[1 + N\ \mathrm{SNR}(f_m)],$$

where $\mathrm{SNR}(f_m)$ is the signal to noise ratio at the frequency f_m. The error covariance matrix becomes

$$\mathcal{E}\{\Delta\underline{D}\Delta\underline{D}^{\mathrm{T}}\} = \frac{1}{2K \sum_{m=m_1}^{m_2} (2\pi f_m)^2 \frac{N\ \mathrm{SNR}^2(f_m)}{1 + N\ \mathrm{SNR}(f_m)}} (I + \mathbf{1}_{N-1,1}\mathbf{1}_{N-1,1}^{\mathrm{T}}). \tag{2.58}$$

For baseband signals and since $\Delta f = K/T$, where T is observation time and K is the number of averages of the periodograms, the righthand side of (2.58) can be rewritten under an integral form as follows

$$\mathcal{E}\{\Delta\underline{D}\Delta\underline{D}^{\mathrm{T}}\} = \frac{1}{T \int_B (2\pi f)^2 \frac{N\ \mathrm{SNR}^2(f)}{1 + N\ \mathrm{SNR}(f)} df} (I + \mathbf{1}_{N-1,1}\mathbf{1}_{N-1,1}{}^{\mathrm{T}}). \tag{2.59}$$

where $B = [-m_2\Delta f, m_2\Delta f]$. The righthand side of (2.59) is the Cramer-Rao lower bound established by Hahn and Tretter [17].

■ *2.3.3. Asymptotic equivalence of the phase unwrapping method and the least square method*

The least square method is based on the maximisation of the criterion (2.52). Asymptotically,

$$\cos\left(\hat{\theta}_n(m) - 2\pi m\Delta f D_n\right) = 1 - \frac{(\hat{\theta}_n(m) - 2\pi m\Delta f D_n)^2}{2}$$

and therefore the maximisation of $\psi(D_n)$ leads to minimising

$$\frac{1}{2}\sum_{m=m_1}^{m_2}(\hat{\theta}_n(m)-2\pi m\Delta f D_n)^2/\text{var}\,\{\hat{\theta}_n(m)\}.$$

The solution is given by

$$\hat{D}_n=\frac{1}{\displaystyle\sum_{m=m_1}^{m_2}\frac{(2\pi m\Delta f)^2}{\text{var}\,\{\hat{\theta}_n(m)\}}}\cdot\sum_{m=m_1}^{m_2}\frac{2\pi m\Delta f\hat{\theta}_n(m)}{\text{var}\,\{\hat{\theta}_n(m)\}}.$$

Assuming that the phase angles $\hat{\theta}_n(m)$, $m_1\leq m\leq m_2$, are statistically independent (which is true if the cross-spectral matrices estimated at different frequency bins are independent), the variance of the time delay estimate $\hat{D}_n$ has the following form

$$\text{var}\{\hat{D}_n\}=\frac{1}{\displaystyle\sum_{m=m_1}^{m_2}\frac{(2\pi m\Delta f)^2}{\text{var}\,\{\hat{\theta}_n(m)\}}}.$$

For orthogonal sources

$$\text{var}\,\{\hat{D}_n\}=\frac{1}{K\displaystyle\sum_{m=m_1}^{m_2}(2\pi f_m)^2\left(\sum_{p=1}^{P}|\,z_p(f_m)\,|^2\,\frac{a(f_m)\lambda_p(f_m)}{\left(a(f_m)-\lambda_p(f_m)\right)^2}\right)^{-1}}\tag{2.60}$$

which is exactly the variance given by (2.57), for the phase unwrapping method.

In the single source case, the variance is given by

$$\text{var}\,\{\hat{D}_n\}=\frac{1}{K\displaystyle\sum_{m=m_1}^{m_2}(2\pi f_m)^2\frac{N\,\text{SNR}^2(f_m)}{1+N\,\text{SNR}(f_m)}},\tag{2.61}$$

which is the variance given by (2.58).

Therefore, for orthogonal sources and a single source, the least square method and the phase unwrapping method lead to the same variance of the time delay estimates. This shows that the phase unwrapping method, for orthogonal sources, treats the time delays separately. Furthermore, in the single source case, both methods reach the Cramer-Rao lower bound for sufficiently large SNR or bandwidth time product.

3. Localisation

This paragraph solves the localisation problem. We recall that the sources and the sensors are located in a d dimensional space ($d = 2, 3$), the origin is taken at the first sensor. Let us consider the differential time delay D_{np} of the pth source on the nth sensor, related to the position $\vec{r}_n$ of the nth sensor and to the normalised bearing $\vec{k}_p$ of the pth source by the formula

$$\vec{r}_n \vec{k}_p = cD_{np}, \quad 1 \leq p \leq P, \quad 2 \leq n \leq N, \tag{3.1}$$

where c is the sound speed in the nondispersive medium. We want to solve the nonlinear system (3.1) with respect to the unknowns $\vec{r}_n$ and $\vec{k}_p$.

Since the time delays D_{np} are given by the inner product of the vectors $\vec{r}_n$ and $\vec{k}_p$, the absolute sensor positions and source bearings cannot be estimated. These parameters are defined up to an orthogonal transformation: if $\vec{r}_n$ and $\vec{k}_p$ are solutions of the system (3.1), and if Q is an orthogonal matrix, the vectors $Q\vec{r}_n$ and $Q\vec{k}_p$ are also solutions of the system (3.1).

The system (3.1) can be rewritten in matrix form. Let us introduce the $d \times (N-1)$ matrix R which defines the sensor positions

$$R = [\vec{r}_2, \ldots, \vec{r}_N]. \tag{3.2}$$

Let K be the $d \times P$ matrix of the source bearings

$$K = [\vec{k}_1, \ldots, \vec{k}_P]. \tag{3.3}$$

And finally let D be the $(N-1) \times P$ matrix of the time delays

$$D = [cD_{np}]. \tag{3.4}$$

Thus, the system (3.1) assumes the form

$$R^{\mathrm{T}} K = D. \tag{3.5}$$

3.1 THE IDENTIFIABILITY CONDITION

In order to solve the localisation problem from the time delays, a minimum number of sources is required. To find this number, let us assume that the time delays have been estimated without any error. The matrix $D = R^{\mathrm{T}} K$, as a product of two matrices of rank smaller or equal to d, has a rank which is smaller or equal to d. If the rank of R and K is exactly equal to d, which is equivalent to the following conditions:

1. The number N of sensors is at least equal to $d+1$ (the first sensor is taken as reference).
2. The number of sources is at least equal to d,
3. The sensors are not aligned in the two-dimensional case, and are not in the same plane in the three-dimensional case,
4. The source bearing vectors spanned $\Re^d$,

then the rank of D is exactly equal to d. Even if we have to renumber the sensors, one can always assume that the d first rows of D are linearly independent. Under these conditions, the d first rows of R^{T} are also independent. Let us define a partition of both matrices R^T and D by isolating their first d rows:

$$R^T = \begin{bmatrix} R_1 \\ \cdots \\ R_2 \end{bmatrix} \begin{matrix} \updownarrow d \quad \text{rows} \\ \\ \\ \end{matrix} \qquad D = \begin{bmatrix} D_1 \\ \cdots \\ D_2 \end{bmatrix} \begin{matrix} \updownarrow d \quad \text{rows} \\ \\ \\ \end{matrix}$$

The rows of D_2 are linear combinations of the rows of D_1, therefore, there exists a rectangular matrix A such that

$$D_2 = AD_1.$$

The initial system can be decomposed as follows:

$$R_1 K = D_1 \tag{3.6}$$

$$R_2 K = D_2 = AD_1. \tag{3.7}$$

If $\{R_1, K\}$ is solution of the system (3.6), $\{AR_1, K\}$ is also solution of the system (3.7). On the order hand, if $\{R_2, K\}$ is solution of (3.7), $\{A^\dagger R_2, K\}$, where $A^\dagger$ is the pseudo-inverse of A, is solution of the system (3.6). This proves that the equations of (3.7) are redundant and that the system (3.6) defines the principal equations of $R^{\mathrm{T}} K = D$. Let $\underline{D}_p$ be the pth column of the matrix D_1

$$D_1 = [\underline{D}_1, \ldots, \underline{D}_P].$$

Since the matrix R_1 is not singular, equation (3.6) is equivalent to

$$K = YD_1,$$

where Y is the inverse of R_1. If the columns of the matrix K are normalised to 1, we get the system

$$\|Y\underline{D}_p\|^2 = 1. \qquad 1 \le p \le P$$

If X is defined as $Y^{\mathrm{T}}Y$, the previous system becomes

$$\underline{D}_p^{\mathrm{T}} X \underline{D}_p = 1, \qquad 1 \le p \le P \tag{3.8}$$

which is a system of P equations with $d(d+1)/2$ unknowns. In order to have a system which has not an infinity of solutions, the following condition on the number of sources must be fulfilled

$$P \geq \frac{d(d+1)}{2}.$$

This defines the necessary condition of identifiability.

Now, let us establish the necessary and sufficient condition for the system (3.8) to have an unique solution. This system can be rewritten in a matrix form. Let us define the $P \times d(d+1)/2$ matrix S constructed with the $d(d+1)/2$ vectors of order P, defined by

$$\underline{s}_{ij} = (2 - \delta_{ij})(D_{i1}.D_{j1}, \ldots, D_{iP}.D_{jP}), \quad 2 \leq i \leq j \leq d+1$$

where δ_{ij} is the Kronecker symbol. Let $\underline{x}_c$ be the column vector of order $d(d+1)/2$ which contains the coefficients x_{ij}, $1 \leq i \leq j \leq d$ of the symmetric matrix X

$$\underline{x}_c = [x_{11}, x_{12}, \ldots, \ldots, x_{1d}, x_{22}, \ldots, x_{2d}, \ldots, x_{dd}]^{\mathrm{T}}.$$

Then the system (3.8) takes the form

$$S\underline{x}_c = \mathbf{1}_{P,1}. \tag{3.9}$$

The matrix equation (3.9) has an unique solution $\underline{x}_c$ if and only if the two following conditions are fulfilled

1. $P \geq d(d+1)/2$

2. The rank of the matrix S is equal to $d(d+1)/2$.

In the two dimensional case ($d = 2$), we show [15] that the two previous conditions can be simply expressed as a condition on the number of sources and their geometrical configuration: the number of sources must be greater or equal to three and three of them must have their bearing vectors not pairwise colinear. Therefore, in the two dimensional case, necessary and sufficient conditions for the solvability of the system (3.5) are the following:

(i) $N \geq 3$ *and the sensors are not on a line*

(ii) $P \geq 3$ *and three of them must have their bearing vectors not pairwise colinear.*

In the three dimensional case, there is no obvious necessary and sufficient condition of identifiability.

3.2 LOCALISATION [18]

In practice, the estimate of the matrix D has a rank equal to P, due to estimation error. In order to decrease the influence of the estimation errors, it is worthwhile to reduce the estimate $\hat{D}$ to a matrix of rank d before solving the system (3.5). This can be achieved by doing the SVD (Singular Value Decomposition) of the matrix $\hat{D}$. Let

$$\hat{D} = U\Sigma V^{\mathrm{T}} = \sum_{i=1}^{\mathrm{P}} \sigma_i u_i v_i^{\mathrm{T}}, \quad \sigma_1 \geq \sigma_2 \geq \ldots \geq \sigma_P, \tag{3.11}$$

be the SVD of $\hat{D}$. Then we replace $\hat{D}$ by the matrix of rank d

$$\tilde{D} = \sum_{i=1}^{\mathrm{d}} \sigma_i u_i v_i^{\mathrm{T}}. \tag{3.12}$$

Once the estimate of the time delay matrix has been reduced to a matrix of rank d, it remains to solve the localisation which is equivalent to solve the equation

$$R^{\mathrm{T}} K = \tilde{D},$$

where $\tilde{D}$ is given by

$$\tilde{D} = U_d \Sigma_d V_d.$$

Σ_d is the diagonal matrix of the d largest singular values of the matrix $\hat{D}$. U_d is the $P \times d$ matrix whose columns are the eigenvectors associated to the d largest eigenvalues of $\hat{D}\hat{D}^{\mathrm{T}}$. V_d is the $d \times P$ matrix whose rows are the eigenvectors associated to the d largest eigenvalues of $\hat{D}^{\mathrm{T}}\hat{D}$. Let us start by the estimation of the source bearing matrix K. As in section 3.1, there exists a positive definite matrix X such that

$$K^{\mathrm{T}} K = V_d^{\mathrm{T}} X V_d$$

and the problem leads to solving the system (3.9) of section 3.1 with P equations and $d(d+1)/2$ unknowns

$$S \underline{x}_c = \mathbf{1}_{P,1},$$

where $\underline{x}_c$ is the column vector which contains the elements of X, $1 \leq i \leq j \leq d$. The matrix S is made up with the $d(d+1)/2$ vectors $\underline{s}_{ij}$ of order P defined by

$$\underline{s}_{ij} = (2 - \delta_{ij})(v_{i1}.v_{j1}, \ldots, v_{iP}v_{jP})^{\mathrm{T}}, \quad 1 \leq i \leq j \leq d,$$

where the v_{kl}, $1 \leq k \leq d$, $1 \leq l \leq P$, are the elements of the matrix V_d.

Assuming that the necessary and sufficient condition of identifiability is fulfilled, the previous system is solved by means of a least square method, in order to find the column vector $\underline{x}_c$. The matrix X, reconstructed from the $\underline{x}_c$, is factored under the form $L^{\mathrm{T}}L$ by the Choleski method, where L is an upper triangular matrix. Then we write

$$K^{\mathrm{T}} K = (LV_d)^{\mathrm{T}}(LV_d),$$

and the matrix K of the normalised source bearings is identified with the matrix LV_d up to an orthogonal transformation. Once the relative source bearings have been estimated, the relative sensor positions are estimated by solving the system, which is now linear,

$$\tilde{D} = K^{\mathrm{T}} R$$

by means of a least square method.

4. Results

The performance of the method will be demonstrated using both simulation and at-sea experimental data. For both cases, the sources and the sensors are located in a plane so that $d = 2$. For the simulation the array is composed of randomly distributed sensors, ensonified by point sources which are transmitting broadband white noise. For the experimental data, the array is linear and three sources transmit broadband signals. In the following, the frequency, the time delays and the sensor positions are normalised and dimensionless. Let f_s be the sampling frequency of the received signals. We define the normalised frequency $\tilde{f}$ and the normalised time delays $\tilde{D}_{np}$ as follow

$$\begin{aligned} \tilde{f} &= f/f_s \\ \tilde{D}_{np} &= f_s.D_{np} \qquad ,2 \leq n \leq N \qquad ,1 \leq p \leq P \end{aligned} \tag{4.1}$$

The sampling frequency f_s is associated with the minimum wavelength $\lambda_{\min} = 2c/f_s$, which is chosen as the length scale and where c is the sound velocity. Then we can define the normalised sensor position by

$$\tilde{\vec{r_n}} = \frac{\vec{r_n}}{(\lambda_{\min}/2)} \quad , \qquad 2 \leq n \leq N. \tag{4.2}$$

In the following, we shall work with these normalised parameters. We start with the simulation results.

4.1 RESULTS OBTAINED BY SIMULATION

The array is composed of 20 sensors distributed in a square whose edge length is 32 (the length scale is $\lambda_{\min}/2$). The positions of the two first sensors are chosen such that $\tilde{\vec{r_1}} = 0$ and $\tilde{\vec{r_2}}$ is colinear to $0\vec{x}$, where $0\vec{x}$ is the unit vector of the x-axis. The array is illuminated by three point sources. The source bearings, from $0\vec{x}$ are equal to 10, 70 and 120 degrees. The three sources transmit independent broadband white noises having the same power spectral density; the spectrum bandwidth B is between 0.2 and 0.4 (the frequency scale is f_s). To take into account the estimation of the cross-spectral matrix, we simulate the time series at the sensor output, not just the cross-spectral matrices at each frequency bin. The source signals are designed as follows. We compute three independent time series of 12 800 samples each, having a zero-mean normal distribution $\mathcal{N}(o, \sigma_s)$, where σ_s is the standard deviation. These three time series are bandpass filtered in the frequency band [0.2, 0.4], correctly delayed and summed up at each sensor. Then we compute 20 independent time series of 12 800 samples each, having a zero-mean normal distribution $\mathcal{N}(o, \sigma_n)$, where σ_n is the standard deviation. These time series represent the uncorrelated additive noises and are summed with the source signals at each sensor.

The cross-spectral matrices $\Gamma(f)$ are estimated, at each frequency, by means of the average of the periodograms:

$$\hat{\Gamma}(f) = \frac{1}{K} \sum_{k=1}^{K} \underline{x}_k \underline{x}_k^*,$$

where $(\underline{x}_k)_{1 \leq k \leq K}$ denotes the K independent complex vectors of order N, obtained by Fourier transform of the observation vector at the array output. The FFT is computed on 128 samples such that the frequency resolution W is equal to 1/128. The number of averages K is equal to 100.

The multifrequency analysis is performed on 20 equispaced frequency bins with a spacing Δf equal to the FFT resolution W. Given these parameters, the bandwidth-time product BT is equal to 4000. Let us now define the signal to noise ratio for each source, at each hydrophone. We assume that the three sources transmit with the same power. Then the signal to noise ratio is given by

$$\mathrm{SNR} = 10 \log_{10} \frac{\int_{0.2}^{0.4} \mid s(f) \mid^2 df}{\int_{0.2}^{0.4} \mid n(f) \mid^2 df}, \tag{4.3}$$

where $s(f)$ and $n(f)$ are respectively the source and noise spectra. Since the spectra are flat in the band, the expression for the signal to noise ratio reduces to

$$\mathrm{SNR} = 10 \log_{10} \frac{\sigma_s^2}{\sigma_n^2}. \tag{4.4}$$

The results presented in this section are obtained with a SNR of -5 dB. The results are presented in Table 1 and Figure 4. In Figure 4, the crosses represent the true sensor positions while the circles represent the estimates. In Table 1, we give the true source bearings and their estimates.

These results demonstrate that the algorithm is capable of accurately reconstructing the array geometry and of accurately finding the source bearings; the maximum error of the bearing estimate is less than 1 degree.

4.2 RESULTS OBTAINED WITH EXPERIMENTAL DATA

The data come from an experiment at sea conducted at SACLANTCEN. A linear array of 30 hydrophones was ensonified by 3 sources located about 12 km away. The sensors are not equispaced; the ten first sensors were closer to each other than the last twenty. The source bearings with respect to the physical axis of the array were approximately 67, 87 and 100 degrees. No assumption was made about the linearity of the array, only the hydrophone spacing was assumed a-priori known. The signals transmitted by the sources were broadband white noises. The multifrequency analysis is performed on the normalised frequency band [0.39, 0.41], using 20 frequency bins, spaced at $\Delta f(\Delta f = 0.001)$. In the frequency band, we define a signal to noise ratio as follows

$$\mathrm{SNR} = 10\log_{10}\frac{\int_{0.39}^{0.41} | s(f) |^2 \, df}{\int_{0.39}^{0.41} | n(f) |^2 \, df}. \tag{4.5}$$

The SNR was approximately equal to 5 dB for each source.

The cross-spectral matrices $\Gamma(f)$ are estimated in each frequency bin by means of an average of short periodograms (FFT of 128 samples), with an overlap of 50%. The number of averages K is equal to 300 and the equivalent number of degrees of freedom is 200. The estimates of the phase vectors at each frequency are obtained by fixing a priori the number of sources to 5, which is the mean value estimated by the MDL criterion over the frequency band [0.39, 0.41]. Figure 5 shows the number of steering vectors estimated at each frequency bin by the monofrequency analysis and the sine of the angle $\alpha(\hat{\underline{\theta}})$ (Eq. 2.3) that they make with the estimated signal subspace. The x-axis is the steering vector number, the y-axis is the frequency in unit of Δf and the z-axis represents $\sin\alpha(\hat{\underline{\theta}})$. We can see that first three steering vectors almost lie in the signal subspace at each frequency (small sine), which suggests the presence of three broadband sources. The results of the multifrequency analysis are shown in Figures 6, 7 and 8. The linkage of the steering vectors shown in Figure 5 estimated 3 broadband sources as we expected from the results of Fig. 5. Figures 6, 7, 8 show, respectively, the absolute values of the phase angle estimates given by

$$\hat{\theta}_{np}(f) = 2\pi f \hat{D}_{np} \quad ,1 \le p \le P, \quad 2 \le n \le N$$

for the three broadband sources. The x-axis is the frequency in units of Δf, the y-axis is the sensor number, and the z-axis is the phase $\hat{\theta}_{np}(f)$. In Figure 7, we clearly see that the phase angle surface is made of two planes with different slopes. The first slope which is small, corresponds to the ten first hydrophones. The second slope, which is larger, corresponds to the last twenty hydrophones. Therefore, at this point of the processing, we already have some a-priori knowledge on the array shape: the array looks piecewise linear.

Let us now discuss the results of the estimation of the source bearings and the sensor positions. The SVD of the time delay matrix estimate $\hat{D}$ (Eq. 3.11) gives one predominant singular value, which means that $\hat{D}$ is of rank one. Therefore one or both of the necessary and sufficient conditions of identifiability (i) and (ii) are not fulfilled. Since we know that the sources are well separated in bearing, i.e. the second condition is satisfied, this means that the array is straight (which is known to be the case); thus we are not able to use the unknown geometry algorithm. Instead we estimate the source directions by solving the system (3.5), where now R contains the known linear sensor positions. The result is shown in Figure 9, where the solid lines represent the true bearings and the dashed lines represent the estimates. The largest error occurs for the source 1; the error on the bearing is approximately 2.5 degrees. For the source 2, the error is around 1.5 degrees. The third source is very well estimated. These results are also subject to small errors due to the experiment measurements of the true bearings.

The experimental results show that the technique presented in this paper can be used with real data. The multifrequency analysis, especially the linkage of the phase vectors, performs well. These first encouraging results with experimental data are still not sufficient. Since the array was straight, it was not possible to demonstrate that the position of the sensors could be estimated. However, it shows that the technique can tell when there is a degenerate situation.

Source Number	True Bearing (Deg)	Estimated Bearing (Deg)
Source 1	10	9.91
Source 2	70	70.15
Source 3	120	120.74

Table 1 Comparison between the true source bearings and their estimates

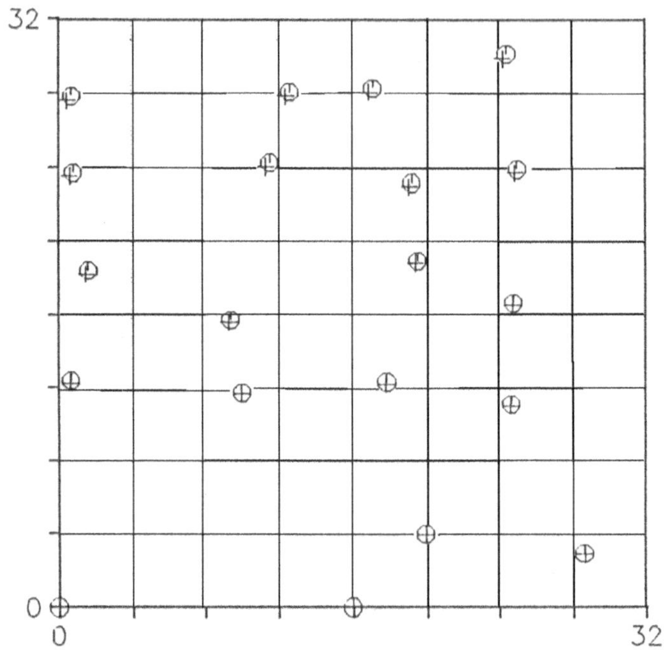

Fig. 4 Comparison between the true sensor positions and their estimates. Number of sensors $N = 20$; number of sources $P = 3$;SNR=-5 dB; $BT = 4000$.

+ True positions

o Estimated positions

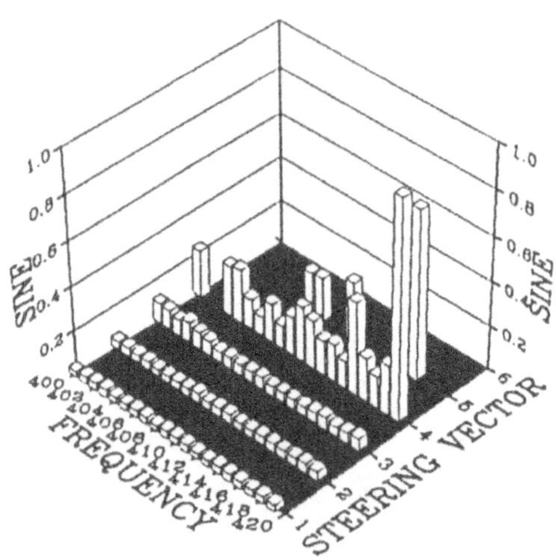

Fig. 5 Sine of the angle between the steering vectors and the signal subspace

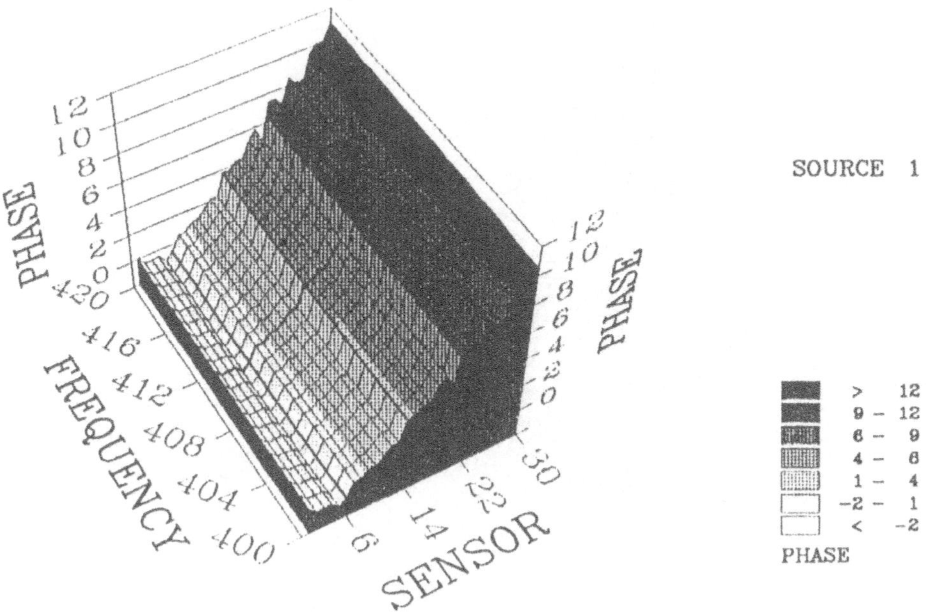

Fig. 6 Absolute phase angle of the source 1

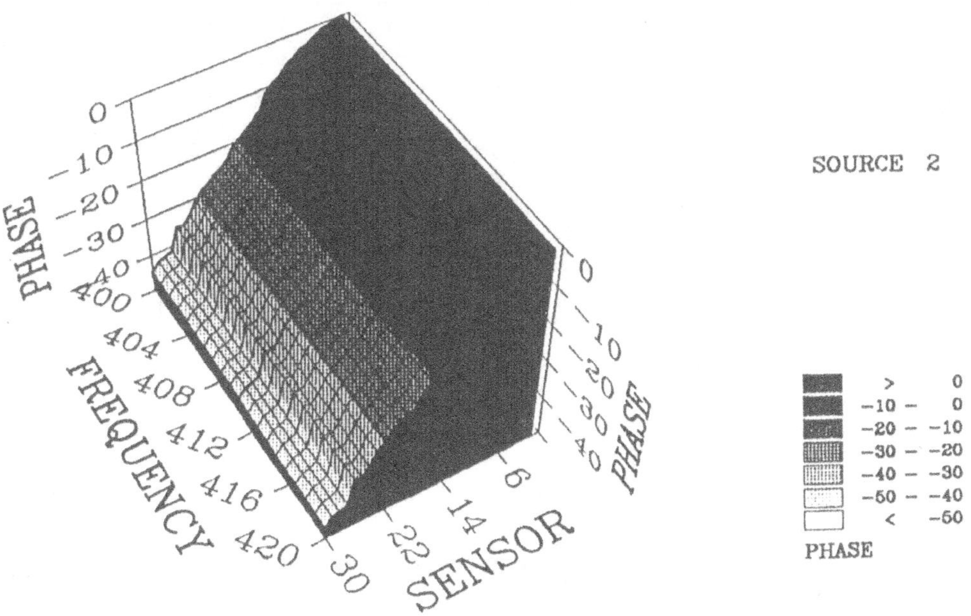

Fig. 7 Absolute phase angle of the source 2

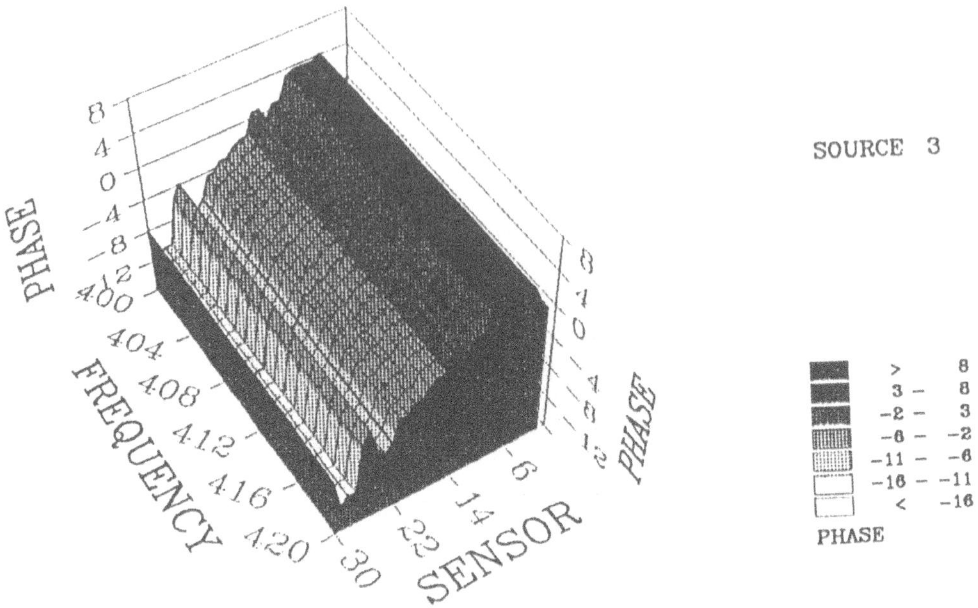

Fig. 8 Absolute phase angle of the source 3

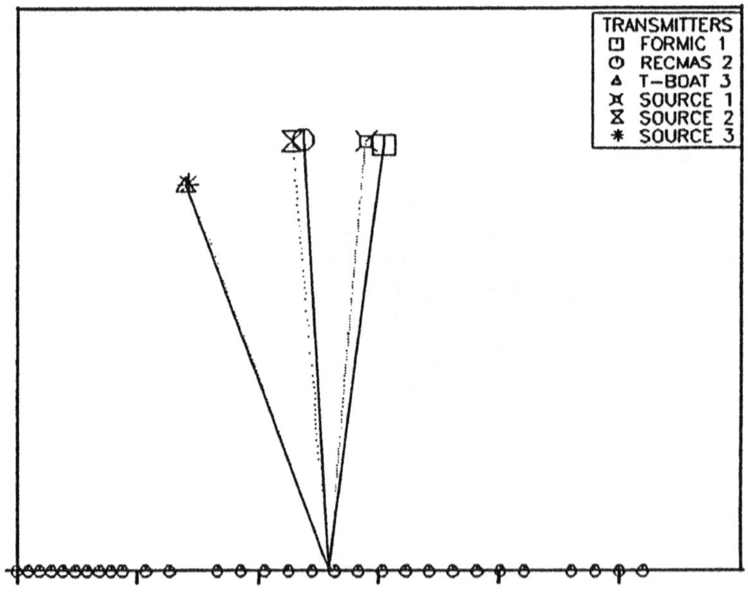

Fig. 9 Estimates of the source bearings

5. Threshold behaviour of the algorithm

This section is devoted to the analysis of the performance of the time delay estimation at low signal to noise ratio (SNR). The performance of the position and bearing estimation are not considered since they involve highly non linear operations (Eq. 3.5). In section 2.3, we saw that the estimation of the differential time delays is efficient at high signal to noise ratio, i.e. the estimates achieve the Cramer-Rao lower bound. However, to fully characterise the time delay estimator, we must analyse the algorithm performance at low SNR, where the estimates are subject to a threshold phenomenon [19]: below a given SNR, the probability of large errors rapidly increases [6] and the Mean Square Error (MSE) of the estimates diverges from the Cramer-Rao bound. Some work has been done to find lower bounds on the MSE which characterises the threshold behaviour among them, the Barankin lower bound [7] and the Ziv-Zakai lower bound (ZZLB) which gives the fundamental limitation in the single parameter estimation [8]. The Ziv-Zakai lower bound has been applied with success to time delay estimation of broadband signals [9] and will be used for our problem. For our method, we know that the time delays are between $[-D_0/2, D_0/2]$ (Eq. 2.42), where $D_0/2 = 1/2\Delta f = D_{max}$ and Δf is the frequency spacing of the multifrequency analysis. In fact, the ZZLB includes explicitly the dependence on the a-priori interval $[-D_0/2, D_0/2]$. For the first investigation of our algorithm performance at low SNR, we will consider the simple case of a single source and two sensors. Before giving the results obtained by Monte-Carlo simulations, let us briefly review the principle of the Ziv-Zakai lower bound. The Ziv-Zakai formulation is based on the probability of deciding correctly between two values D and $D+x$ of the time delay to estimate. Let $P_e(D, D+x)$ be the minimum probability of error associated to the likelihood ratio test, and let us assume that D is uniformly distributed in the a-priori region $[-D_0/2, D_0/2]$. Ziv and Zakai showed that the average mean square error $\bar{e}^2$, which is averaged over $[-D_0/2, D_0/2]$, is bounded as follows

$$\bar{e}^2 \geq \frac{1}{D_0} \int_0^{D_0} x G[(D_0 - x) P_e(x)] dx, \tag{5.1}$$

where $G(x)$ is a non increasing function obtained by filling the valleys in the bracketed function.

Weiss and Weinstein applied this result to time delay estimation [9] and found a lower bound for $P_e(x)$

$$P_e(x) \geq \exp[a(x) + b(x)] \Phi(\sqrt{2b(x)}), \tag{5.2}$$

In the two sensor case, and for large BT and smoothly varying signal and noise spectra, $a(x)$ and $b(x)$ assume the form

$$a(x) = -T \int_0^{\infty} \log[1 + \gamma(f, x)] df, \tag{5.3}$$

$$b(x) = T \int_0^{\infty} \frac{\gamma(f, x)}{1 + \gamma(f, x)} df, \tag{5.4}$$

$$\gamma(f,x) = \frac{s^2(f)\sin^2(\pi f x)}{2s(f)n(f) + n(f)^2}, \tag{5.5}$$

where $s(f)$ and $n(f)$ are respectively the signal and noise spectra. The function $\Phi(y)$ is defined by

$$\Phi(y) = \frac{1}{\sqrt{2\pi}} \int_y^\infty \exp(\frac{-t^2}{2})dt.$$

We test the algorithm for baseband signals defined by

$$\mathrm{SNR}(f) = \begin{cases} \mathrm{SNR} & \mid f \mid \leq \frac{B}{2} \\ 0 & \mid f \mid > \frac{B}{2}, \end{cases} \tag{5.6}$$

where

$$\mathrm{SNR} = \frac{(s/n)^2}{1 + 2s/n}. \tag{5.7}$$

the quantity s/n is the in-band signal to noise ratio at the sensor output. According to (5.5), (5.4) and (5.3), $a(x)$ and $b(x)$ become

$$a(x) = -T \int_0^{\frac{B}{2}} \log(1 + \mathrm{SNR}\sin^2 \pi f x)dx \tag{5.8}$$

$$b(x) = T \int_0^{\frac{B}{2}} \frac{\mathrm{SNR}\sin^2 \pi f x}{1 + \mathrm{SNR}\sin^2 \pi f x} dx. \tag{5.9}$$

For baseband signals, Weinstein and Weiss showed that the Ziv-Zakai lower bound divides the SNR domain in three different segments, as represented in Figure 10.

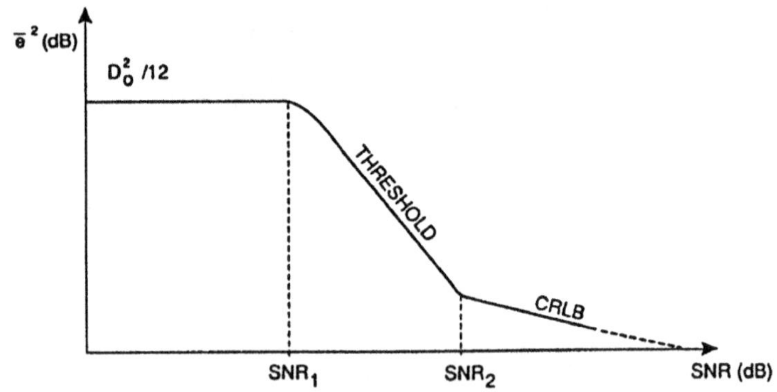

Fig. 10 Ziv-Zakai Lower Bound

If the signal to noise ratio does not exceed SNR_1, the signal at the sensor output is dominated by the noise and is useless for time delay estimation; the MSE is bounded by $D_0^2/12$. If $\mathrm{SNR} > \mathrm{SNR}_2$, the time delay can be estimated without ambiguity and the MSE reaches the CRLB. If $\mathrm{SNR}_1 < \mathrm{SNR} < \mathrm{SNR}_2$ we face the so-called threshold phenomenon, between the ambiguity free-mode and the ambiguity dominated mode. The lower limit SNR_2 of the threshold region depends on the product BD_0.

In order to test the algorithm, we compute the mean square error of the time delay estimates by Montecarlo simulation and compare it to the Ziv-Zakai lower bound. The ZZLB is computed using (5.1) and (5.2) where the quantities $a(x)$ and $b(x)$ are replaced by their expressions (5.8) and (5.9). The baseband signal transmitted by the single source is synthetised as follows. We compute a time series of 12800 samples having a normal distribution $\mathcal{N}(0, \sigma_s)$. This time series is lowpass filtered in the frequency band [0, 0.31]; we use the normalized frequencies defined in section 4. The additive noises at the two sensor outputs are obtained by computing two independent time series of 12800 samples having a normal distribution $\mathcal{N}(0, \sigma_n)$. These two time series are lowpass filtered in the same frequency band, and added to the signal. The signal to noise ratio at the sensor output is defined by

$$s/n = 10 \log_{10} \frac{\sigma_s^2}{\sigma_n^2}. \tag{5.10}$$

The cross-spectral matrices $\Gamma(f)$ are estimated by means of the average of short periodograms (FFT of 128 points). The number K of averages is 100, the frequency spacing Δf is equal to 1/128, and the a-priori region of the time delays is [-128, 128]; the time scale for time delays has been defined in section 4. The bandwidth time product BT is equal to 4000 and BD_0 is equal to 40. The number of sources is a-priori fixed to one and the steering vector associated to the signal source is given by the largest eigenvector of $\hat{\Gamma}(f)$. In this case, the linkage algorithm is not used.

For a given signal to noise ratio, the MSE is computed from 100 independent runs. Figure 11 shows the results. The x-axis is the signal to noise ratio defined by (5.7) and the y-axis is the MSE normalised by $D_0^2/12$ and given in dB. The solid line represents the estimated MSE, the dashed line represents the ZZLB and the dotted line represents the CRLB. This result points out the three SNR domains with the threshold effect and is very encouraging since the MSE of the estimated time delays is close to the ZZLB within 1 dB.

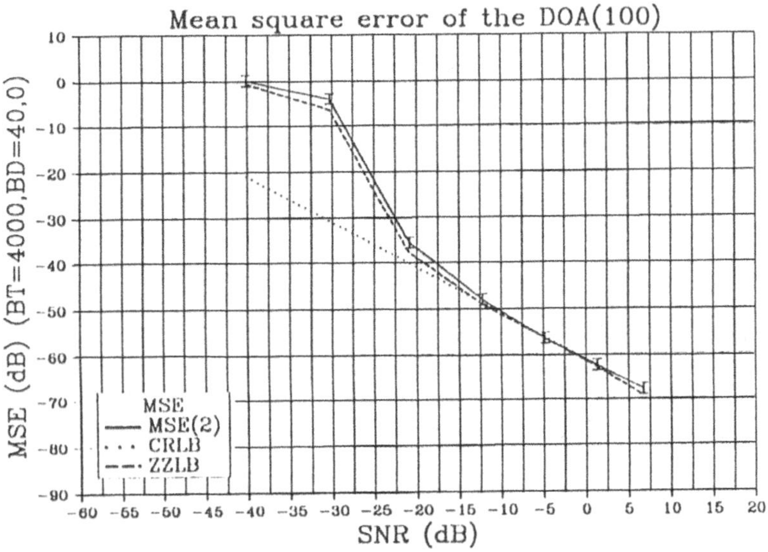

Fig. 11 MSE of the time delay estimate

6. Conclusions

The present work shows the feasibility of the localisation of multiple broadband sources with an array of unknown geometry under the following conditions:

- The number of sensors is much greater than the number of sources;
- A minimum number of sources is required; three sources in the two dimensional case and six sources in the three dimensional case;
- There is no differential attenuation of each coherent wavefront across the array.

The algorithm allows us to reconstruct the relative array geometry and to estimate the relative source bearings simultaneously, in coordinate axes linked to the array. The determination of the absolute sensor positions and the absolute bearings is possible if we know the exact positions of three sensors in the two dimensional case and four sensors in the three dimensional case.

The results obtained by simulations quantify the validity of the method. When the array is composed of twenty sensors randomly distributed in a square of edge 32 $\lambda/2$ and insonified by three broadband sources, the error on the bearing estimate is less than one degree. The in-band signal to noise ratio at the sensor outputs is equal to -5 dB. Since the time series are computed at the sensor outputs, these simulations take into account the estimation procedure of the cross-spectral matrix and consequently the bias of the estimates. Like all the eigenvector techniques, the method requires a good estimation of the cross-spectral matrix with a bias as small as possible.

The results obtained with at-sea experimental data show that the method presented in this paper can be used with real data. The multifrequency analysis, especially the linkage of the phase vectors, performs well. These first encouraging results with experimental data are still not sufficient. Since the array was straight, it was not possible to demonstrate that the positions of the sensors could be recovered. However, it shows that the technique can tell when there is a degenerate situation. Therefore, a further test with a field of randomly distributed sensors, which is in fact the intended application of the algorithm, is needed to demonstrate the validity of the method.

The first investigation of the algorithm performance at low SNR, for the simple case of a single source and two sensors, exhibits the threshold phenomenon. However, the mean square error of the time delay estimate comes very close to the Ziv-Zakai lower bound, which demonstrates good threshold bahaviour for the algorithm. The next step of the performance analysis will be for a single source and N linear equispaced sensors. In that case, the $N-1$ differential time delays will be transformed into bearing and the mean square error of the estimate will be compared to the ZZLB [9]. A more complicated situation is a single source with N randomly distributed sensors. The ZZLB cannot be used anymore as a lower bound on the MSE of the time delay vector, since it is a single parameter bound. At this point, a new bound needs to be derived, which handles multidimensional parameters. This analysis of performance is important to validate the algorithm but is also interesting in term of array geometry. It is known that a linear array is not the optimum design for low MSE of time delay estimates. An array of randomly distributed sensors might improved the performance of the time delay estimation and consequently the localisation.

References

[1] J. Capon, " Maximum likelihood spectral estimation. Non-linear methods of spectral analysis," S.Haykin, editor. Berlin Springer, 1983, pp. 155-179.

[2] B. Porat and B. Friedlander, "Estimation of spatial and spectral parameters of multiple sources," IEEE Trans. Inform. Theory, vol. IT-29, pp. 412-425, May 1983.

[3] W.S. Ligget, "Passive sonar: fitting models to multiple time series," in Signal Processing, J.W.R. Griffiths, P.L. Stocklin, and C. Van Schooneveld, editors. Academic Press 1973, pp.327-345.

[4] H. Mermoz, "Imagerie, correlation et modeles", Annales des Telecommunications, vol. 31, pp. 17-36, Janvier 1976.

[5] G. Bienvenu and L. Kopp, "Source power estimation method associated with high resolution bearing estimator," in Proc. ICASSP 81, Atlanta, Georgia, GA, April 1981, pp. 153-156.

[6] J.P. Ianniello, "Time delay estimation via cross-correlation in the presence of large estimation errors," IEEE Trans. Acoust., Speech, Signal Processing, Vol. ASSP-30, pp. 998-1003, Dec. 1982.

[7] R.J. Mc Aulay and L.P. Seidman, "A useful form of the Barakin lower bound and its application to PPM threshold analysis," IEEE Trans. Inform. Theory, Vol. IT-15, pp.273-279, March 1969.

[8] D. Chazan, M. Zakai and J. Ziv, "Improved lower bounds on signal parameter estimation," IEEE Trans. Inform. Theory, Vol. IT-21, pp. 90-93, January 1975.

[9] E. Weinstein and A.J. Weiss, "Fundamental limitations in passive time delay estimation-Part II:-Wideband systems ," IEEE Trans. Acoust., Speech, Signal Processing, Vol. ASSP-32, pp. 1064-1078, Oct. 1984.

[10] G. Vezzosi and P. Nicolas, "Separation de front d'ondes correles," in Proc. GRETSI 1983, Nice, France , May 1983, pp. 277-282.

[11] M. Wax and T. Kailath, "Determining the number of signals by information theoretic criteria," in Proc. ASSP Spectr. Estimat. Workshop II, Tampa, FL, Nov. 1983, pp. 192-196.

[12] G. Vezzosi, "Estimation of phase angles from the cross spectral matrix," IEEE Trans. Acoust.,Speech, Signal Processing, vol. ASSP-34, pp. 405-422, June 1986.

[13] D.G. Luenberger, Introduction to Linear and Non-linear Programming. Reading, MA, Addison-Wesley, 1973.

[14] D.R. Brillinger, Time Series, Data Analysis and Theory. San Francisco, CA: Holden Day, 1981.

[15] P. Nicolas, "Localisation de sources ponctuelles avec une antenne de geometrie inconnue", These de 3e cycle, Universite' de Rennes I, France, Sept. 1985.

[16] J.M. Tribolet, "A new phase unwrapping," IEEE Trans. Acoust., Speech, Signal Processing, Vol. ASSP-25, pp. 170-177, April 1977

[17] W.R. Hahn and S.A. Tretter, "Optimum processing for delay vector estimation in passive sonar arrays," IEEE Trans. Inform. Theory, vol. IT 19, pp.608-614, Sept. 1973.

[18] W. Brandenburg, "Measurement of Azimuth and Distance using Arbitrary Sensor Configuration with Unknown Coordinates," in Proc. ICASSP 82, Paris, 3-5 May 1982, pp. 1120-1123.

[19] S. K. Chow and P.M. Schultheiss, "Delay Estimation Using Narrowband Processes", IEEE Trans. Acoust., Speech, Signal Processing, Vol. ASSP-29, No. 3, pp. 478-484, June 1981.

CHAPTER 4

HIGH-RESOLUTION PROCESSING TECHNIQUES FOR TEMPORAL AND SPATIAL SIGNALS

H.J. WHITEHOUSE
Naval Ocean Systems Center, San Diego (USA)

B. BOASHASH
University of Queensland (Australie)

P.M. SPEISER
Naval Ocean Systems Center, San Diego (USA)

Chapter 4

High Resolution Processing Techniques for Temporal and Spatial Signals

H. J. Whitehouse[1]
B. Boashash[2]
J. M. Speiser[3]

4.1 Introduction

The first modern generation of signal processing algorithms used Fourier methods. Their theoretical basis was provided by the convolutions which occur in time-invariant linear systems and the important role of sinusoids as eigenfunctions of every such system. A second class of signal analysis and parameter estimation methods use a general linear model similar to the one used in the statistical literature. Representative methods in this class are the linear predictive and signal subspace methods for parameter estimation. A third class of signal processing methods based on time-frequency analysis of nonstationary signals began with the work of Wigner (1932) and

[1] Naval Ocean Systems Center, San Diego, CA 92152-5000
[2] CRISSP, University of Queensland, Brisbane 4067, Australia
[3] Naval Ocean Systems Center, San Diego, CA 92152-5000

Ville (1948) and has seen extensive theoretical development and the beginning of hardware implementation in the 1978-1988 period. Recent work has combined linear predictive modelling with Wigner-Ville analysis. The near future will see the combination of time-frequency distribution (TFD) methods with both eigensystem-based signal analysis and wideband ambiguity function and wideband WVD analysis in which Doppler is treated as a time compression or expansion.

This paper provides an updated discussion of high resolution processing techniques for temporal and spatial signal processing, with particular emphasis on methods applicable to underwater acoustic signal processing. The paper consists of three parts: The first part treats direction finding and spectral estimation for stationary random processes. The second part discusses time-frequency analysis for nonstationary signals using both the Wigner-Ville distribution (WVD) and an autoregressive modified WVD. The third part discusses the application of these techniques to radar and sonar imaging.

4.2 Direction Finding

This section provides an overview of selected beamforming and direction finding (df) techniques and their computational requirements. It provides a description of algorithms and an assessment of their range of applicability. Last it recommends areas for future study. With a few noted exceptions, it considers only narrowband processing - i.e. it is assumed that the data snapshots are the Fourier coefficients in one frequency bin for each of the array elements for the currrent observation interval, and that different frequency bins are processed independently.

4.2.1 Algorithms Considered

- Classical Beamforming
- MVDR and Capon's Method
- DF by Linear Predictive Spectral Estimation
- MUSIC
- Johnson & DeGraf
- Other MUSIC variants
- Pisarenko Harmonic Retrieval

- ESPRIT
- ESPRIT variants
- Cadzow's Eigenvector Approximation Method
- Maximum Likelihood

4.2.2 Principal Notation for Direction Finding

$\mathbf{x}$	data vector
$\mathbf{x}^{(k)}$	data vector for kth snapshot
K	total number of snapshots currently available
M	number of elements in array
B	number of beams
W	total processed bandwidth
a(u)	arrival phase shift vector for look direction u. As a function of u, also called the "array manifold"
A	1) general matrix 2) steering phase shift matrix $A = (\mathbf{a}(\mathbf{u}_1) \cdots \mathbf{a}(\mathbf{u}_D))$
D	1) number of arriving wavefronts (can be larger than the number of sources) when multipath arrivals are considered) 2) diagonal matrix in svd or eigenvalue decomposition
s	steering vector. alternate notation for a(u)
E	usually denotes statistical expectation
H	used as a superscript to denote Hermitian transpose
P	1) source correlation matrix 2) unitary matrix
Q	1) unitary matrix 2) matrix with orthonormal columns - part of a unitary matrix
U	usually used to denote upper (or right) triangular matrix. In the numerical analysis literature, R is usually used to denote a right triangular matrix. However, since the signal processing literature usually uses R for a correlation matrix, we will usually use U instead, except in the names of numerical algorithms, such as QR decomposition and the QR eigensystem method.
R	spatial correlation matrix. $R = E\mathbf{x}\mathbf{x}^H$ When it is desired to emphasize the dependence on $\mathbf{x}$, we will use $R_{\mathbf{x}}$
R_n	1) noise spatial correlation matrix 2) scaled noise correlation matrix
w	weight vector (combined amplitude and phase)

PDQ^H	factors in singular value decomposition. P is unitary D is diagonal, and Q is unitary
QDQ^H	factors in eigenvalue decomposition of a Hermitian matrix. Q is unitary, D is diagonal.
n	noise vector
f	1) vector of complex amplitudes for wavefronts arriving at the array 2) frequency
σ^2	noise power. possibly after spatially prewhitening

4.3 Overview and Algorithm Assessment

4.3.1 Classical Beamforming

We consider classical beamforming only to establish notation and terminology that will be used in discussing additional algorithms. Standard references for classical beamforming include Ma [30] and Steinberg [65]. For most algorithms, we will regard the frequency-domain output of the current data snapshot for the array as a single complex vector, **x**. The data model used is:

$$\mathbf{x} = A\mathbf{f} + \mathbf{n}$$

Classical frequency domain beamforming via phase shift and sum is the frequency domain equivalent of delay and sum time domain beamforming. For each look direction u, it forms the inner product of the steering phase shift vector s = a(u) with the data vector **x**. This inner product, b = [a(u),x] is the spatial equivalent of a matched filter. A derivation for general array geometries with simplifications for special array geometries has been provided by Speiser, Whitehouse, and Berg [59]. The function a(u) is sometimes called the "array manifold" [52].

Shading may be used to reduce sidelobe height, at the expense of reduced resolution. Special array geometries permit the beamforming to be performed via an FFT. These include uniform line and planar arrays, circular and cylindrical arrays, and logarithmic linear and spiral arrays [59].

4.3.2 MVDR Beamforming

Minimum Variance Distortionless Response (MVDR) beamforming attempts to reject interfering sources while maintaining unit gain and zero phase shift for each look direction [40] [43].It is sometimes referred to as "Optimum Array Processing" [40]. Unlike classical shaded beamforming, it does not try to create uniformly low sidelobes in the response pattern, but only to form nulls in the direction of interfering sources. For each steering vector $\mathbf{s} = \mathbf{a}(\mathbf{u})$, it forms $b = [\mathbf{w}, \mathbf{x}]$, where the weight vector $\mathbf{w}$ is chosen to minimize the expected power output, while providing unit gain and zero phase shift for a signal arriving from direction $\mathbf{u}$. i.e. pick $\mathbf{w}$ to minimize $E|b|^2$ subject to $[\mathbf{w},\mathbf{s}] = 1$.

$$\mathbf{w}_{opt} = \frac{R^{-1}\mathbf{s}}{[R^{-1}\mathbf{s}, \mathbf{s}]}$$

MVDR is not a high resolution method per se, but it is recommended as a baseline for comparison. It is quite robust, useful with any geometry, and very well understood, both in terms of its performance and parallel algorithms and architectures for real time implementation, including variants incorporating multiple linear constraints [54] [32].

Apart from resolution MVDR does have one limitation: coherent multipath arrivals in the same frequency bin may cause it to effectively lose sensitivity. That is, when the beam is steered to one of the arrivals, even though the weight vector provides unit gain in that direction, the response to a coherent multipath arrival may reduce the signal component of the output.

4.3.3 Capon's Method

Capon's method estimates the angular spectrum, or power versus arrival direction as $1/[R^{-1}\mathbf{s}, \mathbf{s}]$ [17], [18], [26]. This is the average power output of an MVDR beamformer with steering vector $\mathbf{s}$. The signal processing literature (including Schmidt's papers) frequently refers to Capon's method as "maximum likelihood". This is a source of confusion, since it does not perform maximum likelihood parameter estimation, and does not have the high resolution of true maximum likelihood spectrum estimation/direction finding. We will regard Capon Maximum Likelihood as just an auxiliary output easily available from an MVDR beamformer.

4.3.4 DF by Linear Predictive Spectral Estimation

Linear predictive spectral estimation determines the weights of a prediction error filter for a wide-sense stationary random sequence.

Such a filter converts the sequence into white noise. Since the spectral density of the output of a filter is the product of the input spectral density with the magnitude squared of the filter's transfer function, the input spectral density function is proportional to the reciprocal of the magnitude squared of the filter's transfer function. When the input process has an all-pole spectrum, the optimum prediction based on the infinite past is the same as the optimum predictor using a number of past samples equal to the number of poles. This method is sometimes called Maximum Entropy Spectral Estimation - more specifically the Yule-Walker version of Maximum Entropy Spectral Estimation [26],[35]. It has been proposed as a direction finding technique by utilizing the equivalence of spectral estimation for a wide-sense stationary random process with power versus arrival direction estimation for a uniformly spaced line array receiving signals from uncorrelated sources [31]. Linear predictive spectral estimation can provide high resolution, but require special precautions to avoid line splitting and bias errors [35]. The optimization of a signal processing function analogous to a thermodynamic function can be used not only to derive linear predictive estimators, but also can be used in a new class of estimators which can be used with a small number of data snapshots [56].

4.3.5 MUSIC

The MUSIC method is at this time the most thoroughly studied and best understood of the eigensystem based direction finding techniques. It was independently developed by R. Schmidt [52], [53] in the U. S. and by the team of L. Kopp and G. Bienvenu in France [6]. Schmidt's formulation uses the generalized eigensystem of the data correlation matrix $R_{\mathbf{X}}$ and the scaled noise alone spatial correlation matrix $R_{\mathbf{n}}$. The Bienvenu and Kopp formulation (and many others who have studied MUSIC) assume that the noise is spatially white or the data has been spatially prewhitened, so that $R_{\mathbf{n}} = \sigma^2 I$. The noise power, σ^2 is regarded as an unknown parameter. In this case, the generalized eigensystem reduces to the ordinary eigensystem of $R_{\mathbf{X}}$ alone. In discussing MUSIC it is necessary to distinguish between the behavior with perfect knowledge of the data correlation matrix and scaled noise alone correlation matrix, versus behavior with estimated or modelled correlation matrices.

MUSIC requires several assumptions:

1. The number of arriving wavefronts (in the frequency bin being analyzed) is strictly less than the number of elements in the array.

2. The columns of the arrival phase shift matrix, Λ, are linearly independent.

3. The source correlation matrix, P, is strictly positive definite. (sources can be coherent - but not perfectly coherent)

4. The noise spatial correlation matrix is nonsingular. (This requirement is unlikely to ever cause a problem)

With perfect knowledge of the data correlation matrix and scaled noise spatial correlation matrix, MUSIC proceeds as follows:

Solve the generalized eigensystem

$$R_{\mathbf{x}}\mathbf{z} = \lambda R_{\mathbf{n}}\mathbf{z}$$

If the generalized eigenvalues are numbered in decreasing order, the last $M - D$ are equal, and the first D generalized eigenvalues are strictly greater. The span of the generalized eigenvectors corresponding to the last $M - D$ generalized eigenvalues is called the noise subspace. Each vector in the noise subspace is orthogonal to the arrival phase shift vector $\mathbf{a}(\mathbf{u}_d)$ from each source direction. The number of sources is estimated as $D1 = M - M_0$ where M_0 is the multiplicity of the smallest generalized eigenvalue.

The arrival directions are then determined by looking for array manifold vectors $\mathbf{a}(\mathbf{u})$, orthogonal to the noise subspace. This determines the arrival phase shift matrix A. Then the source correlation matrix P, can be computed as

$$P = (A^H A)^{-1} A^H (R_{\mathbf{x}} - \lambda_{min} R_{\mathbf{n}}) A (A^H A)^{-1}.$$

Of course it is preferable to use the pseudovinverse of A, computed by an SVD, rather than forming

$$(A^H A)^{-1} A^H$$

in order to avoid problems of ill-conditioning caused by near collinearity of columns of A when signals arrive from nearly the same direction.

Early simulations by Schmidt, using perfect correlation matrices, showed that MUSIC had far higher resolution than classical beamforming and Capon's maximum likelihood method, and did not have the bias problems of the maximum entropy method.

It is important to note that the MUSIC "DOA spectrum" is simply a function that peaks in directions where the steering vector is nearly orthogonal to the noise subspace. That is, peaks correspond to arrival directions, but the DOA spectrum is not a measure of arrival power versus direction. Furthermore, the processing is highly nonlinear, so there is no simple interpretation of the apparent peak-to-sidelobe ratio.

The previous comments must be modified somewhat when estimated correlation matrices are used, but MUSIC still remains a very strong contender for direction finding. When estimated correlation matrices are used:

1. Instead of a smallest generalized eigenvalue with multiplicity (M - D), there is a spread of small generalized eigenvalues. If we pick too large a dimension for the noise subspace, then signals will be missed. If we pick somewhat too small a dimension for the noise subspace,

the method degrades gracefully: Spurious peaks will be found in the DOA spectrum, but these can be eliminated through estimation of the source correlation matrix and eliminating peaks corresponding to zero power.

2. With ideal data and scaled noise correlation matrices, MUSIC can localize in the presence of highly correlated (but not perfectly correlated) multipaths. In simulations using estimated correlation matrices, problems have been reported when the correlation between multipath arrivals was around .7 or so.

A survey of work on MUSIC and related eigenvector techniques through 1985, with detailed discussion of source covariance matrix estimation and a signal estimation techniques is available [61].

4.3.6 Johnson & DeGraaf

The method of Johnson and DeGraaf is a slight modification of MUSIC. In MUSIC, the generalized eigenvalues are only used in deciding on the dimensionality of the noise subspace, not in forming the DOA spectrum - i.e. all the basis vectors for the noise subspace are weighted equally in the DOA spectrum. Johnson and DeGraaf use a modified DOA spectrum, in which the eigenvector weighting depends on the corresponding eigenvalue, so that in effect the decision as to the dimension of the noise subspace (or equivalently the number of signals present) is a soft decision [27].

In simulations by the inventors, it was found to provide the same performance as MUSIC when the correct number of signals was chosen for both, but to be more robust when errors were made in picking the number of signals. Studies by Gordon Martin confirmed this behaviour when simulated data was used, but showed better performance for MUSIC when actual HF electromagnetic data was used [36].

4.3.7 Other MUSIC variants

Two MUSIC variants intended to permit use with highly correlated multipath arrivals are applicable for uniformly spaced line arrays. One technique uses averaging of the correlation matrices from subarrays [55], [49]. In effect this tends to restore the Toeplitz structure that one would expect with uncorrelated sources. However, it both reduces the effective aperture and the number of arrivals that can be handled by a factor equal to the number of subarrays used. The second technique is called Root MUSIC. For a uniform line array, the steering vectors are sinusoids, so finding steering vectors orthogonal to the noise subspace is equivalent to finding zeros of polynomials on the unit circle. In Root MUSIC, zeros are rejected if they lie too far from the unit circle, and zeros close to the unit circle are projected onto the unit circle [50]. Root MUSIC is reported to provide better threshold performance than classical MUSIC.

4.3.8 Pisarenko Harmonic Retrieval

Two spectral estimation techniques are both sometimes referred to as "Pisarenko's Method". One is a family of methods based on the eigenvalues and eigenvectors of the correlation matrix. Particular members of this family are chosen by picking a monotone function and its inverse. In this report, we discuss only the Pisarenko special harmonic retrieval method [48]. This method determines the frequencies and amplitudes of sinusoids in additive white noise. Although it preceded (and inspired) MUSIC, it is best understood as a special case of MUSIC, with the following additional assumptions:

1. The noise is white.
2. The data correlation matrix is Toeplitz. In df application this corresponds to uncorrelated sources and a uniformly spaced line array.
3. The noise subspace is treated as having dimension equal to 1.

The Pisarenko method requires finding the eigenvector of a Toeplitz correlation matrix corresponding to the smallest eigenvalue, and then finding sinusoid vectors orthogonal to that eigenvector. Although it can provide high resolution it is not recommended for the following reasons:

1. The Toeplitz assumption makes it inapplicable when even partially coherent multipaths are present.
2. Simulation studies and theoretical analysis have shown that the frequency (or angle in the df application) estimates have large variance or poor statistical efficiency.
3. Gordon Martin's studies using HF data have confirmed the large variance of the direction estimates [36].

4.3.9 ESPRIT

ESPRIT requires an array consisting of two identical subarrays, with the same displacement vector between each element of the first array and the corresponding element of the second array. Alternatively, the total set of array elements may be regarded as a set of M dipoles, with the same differential displacement vector (length an orientation) for each dipole. The following advantages are claimed for ESPRIT in comparison with MUSIC [45], [46], [47], [50]:

1. ESPRIT eliminates the computationally expensive search through the DOA spectrum.

2. It provides better performance in the presence of highly correlated multipaths.

3. It provides a lower noise threshold.

4. It does not require knowledge of the array manifold.

However, the following disadvantages of ESPRIT for spatial signal processing should also be noted [63]:

1. It is applicable only to a very limited set of array geometries.

2. Even when a two-dimensional array is used, it only produces the same information as other methods using a line array - i.e. arrival angle with respect to the differential displacement axis.

3. If the length of the differential displacement vector exceeds half a wavelength, then ESPRIT will have grating lobe ambiguities.

For applications other than spectrum analysis and direction finding with a uniformly spaced line array spaced closer than spatial Nyquist, the above problems limit the usefulness of ESPRIT.

4.3.10 ESPRIT Variants

The original formulation of ESPRIT was based on solving a non-symmetric generalized eigensystem. Two recent alternatives have been proposed: Total Least Squares ESPRIT, and Orthogonal Procrustes ESPRIT [78],[79],[80], [51]. Total Least Squares ESPRIT is claimed to provide slightly better performance at low signal to noise ratio.

4.3.11 Cadzow's Eigenvector fitting method

Cadzow's method is a modification of MUSIC that avoids the requirement that the source correlation matrix be strictly positive definite [16]. That is, perfectly coherent sources are permitted. The method starts with the eigensystem and "DOA spectrum" calculations as in MUSIC. The noise is assumed to be spatially white, or prewhitened if necessary. Then it is observed that each eigenvector of the data correlation matrix is a linear combination of steering vectors from the source directions. Source directions are then found by fitting each eigenvector by linear combinations of steering vectors. Usually only a few steering vectors (corresponding to highly correlated arrivals) are used to fit each eigenvector. Since the steering vectors depend nonlinearly on the arrival directions, this is a nonlinear least squares problem, and must be solved iteratively.

4.3.12 Maximum Likelihood Parameter Estimation

This technique (not to be confused with Capon's so-called "maximum likelihood" method) is based on the statistical maximum likelihood parameter estimation method, and assumes that the noise distribution is Gaussian. Maximizing the likelihood ratio for jointly estimating the signal amplitudes and arrival directions is then equivalent to a nonlinear least squares fit of the data vectors by linear combinations of steering vectors - a computational task similar to that required by Cadzow's method. It is known from the statistical literature that maximum likelihood parameter estimation has excellent large sample properties - asymptotically unbiased, consistent, and asymptotically efficient. Recent empirical studies of maximum likelihood direction finding suggest that the small sample behavior is very competitive with other high resolution techniques, and that the method is robust in the presence of highly coherent sources [14], [72], [77]. It also extends in a straightforward way to broadband processing.

4.4 Introduction to Nonstationary Signal Analysis using the Wigner-Ville Distribution

An increasing set of signal processing tasks require an extension of spectral analysis to treat nonstationary functions of a time or space variable. Traditional techniques for the analysis of signals and stationary random processes are based on the fact that the sinusoids are eigenfunctions of every shift-invariant linear operator. This includes both the impulse response of a time-invariant linear filter and the covariance function of a wide-sense stationary random process. The Karhunen-Loeve (K-L) expansion represents an arbitrary random process in terms of the eigenfunctions of its covariance function, and therefore permits an expansion in terms of uncorrelated random variables. However, the K-L expansion does not posess convenient properties for the analysis of signals that have been filtered or modulated.

Recent signal analysis techniques using time-frequency distribution (TFD) functions provide a bridge between the the traditional stationary signal analysis methods and the yet to be developed methods for treating general nonstationary problems. The Cohen [19] class of time-frequency distributions provide a time-varying spectrum with many useful properties for a wide class of signals. Different TFDs in the Cohen class are selected via the choice of a weight function. The choice of a weight function determines which useful properties will be posessed by the resulting TFD, and no one choice provides all of the possible desirable properties.

4.4.1 Principal Notation for Time-Frequency Analysis

t	time
f	frequency
i	imaginary unit (also used as an index)
$s(t)$	real signal
$\mathcal{H}$	Hilbert transform
$\hat{s}(t)$	alternate notation for the Hilbert transform of $s(t)$
A^H	Hermitian transpose of a matrix A
$z(t)$	complex analytic signal
$W_z(t,f)$	Wigner-Ville distribution of $z(t)$
$S_z(t,f)$	short-time Fourier transform of $z(t)$
$K_z(t,\tau)$	bilinear kernel corresponding to $z(t)$
$\star_f$	convolution with respect to frequency
$\star_t$	convolution with respect to time
$\star_{tf}$	two-dimensional convolution in t, f
$m(t)$	window or modulation function
f_0	initial frequency
f_c	carrier frequency
f_s	sampling frequency
α	chirp parameter
T	length of analysis window
$\epsilon(n)$	white noise sequence
$c(i,j)$	estimated covariance matrix
$a(i)$	autoregressive model parameters
C^+	Moore-Penrose pseudoinverse of matrix C
λ	eigenvalue
$\mathbf{v}$	eigenvector
σ	standard deviation (also used for singular value)

4.4.2 The Wigner-Ville Distribution

One TFD that has a large set of the possible useful properties, and that has received extensive study in recent years for application to signal processing is the Wigner-Ville Distribution (WVD) [73],[37]. The WVD of a real signal $s(t)$ that has complex analytic signal $z(t)$ is defined as

$$W_z(t,f) = \int z(t+\tau/2)z^*(t-\tau/2)e^{-i2\pi f\tau}\,d\tau$$

where $z(t) = s(t) + i\hat{s}(t)$ and $\hat{s}(t)$ denotes the Hilbert transform of $s(t)$. The WVD can also be expressed as

$$W_z(t,f) = \int K_z(t,\tau)e^{-i2\pi f\tau}\,d\tau$$

where the kernel $K_z(t,\tau)$ is defined as

$$K_z(t,\tau) = z(t+\tau/2)z^*(t-\tau/2).$$

The WVD of a chirp (a function with constant amplitude and quadratic phase) is distributed on a line through the origin of the time-frequency plane, with slope proportional to the chirp rate. The WVD of the convolution of two signals is the convolution in time of their respective WVDs:

$$W_{z*h}(t,f) = \int W_z(t-u,f)W_h(u,f)\,du$$

. These two properties are crucial to a number of applications of the WVD:

- Location of Sources in the Fresnel Region
- Coherent Integration of Lines Emitted by a Moving Source
- Identification of Stationarity Intervals for a Nonstationary Random Process
- Radar Imaging

Since a narrowband signal or signal component in the Fresnel region of a line array produces a quadratic phase shift as a function of position along the array, such a source can be localized in range and bearing via a spatial WVD in the temporal Fourier transform domain. [13].

Estimation techniques using signal subspace concepts can be applied to time-frequency distributions by replacing the Fourier transform of the bilinear kernel by a parametric modelling with respect to the variable τ at each time t [8], [12]. Such a parametric high-resolution modified WVD can be used to enhance the resolution attainable in the spatial localization of a source in the Fresnel region of a line array [74], [66].

4.5 The Windowed Wigner-Ville Distribution

Let s(t) be a real continuous-time signal. The analytic signal associated with s(t) may be expressed as

$$z(t) = s(t) + i\mathcal{H}[s(t)] \tag{4.1}$$

Where $\mathcal{H}$ denotes the Hilbert Transform.

The WVD of a signal s(t) is defined as follows:

$$W_z(t,f) = \int_{-\infty}^{+\infty} z(t+\tau/2)z^*(t-\tau/2)e^{-j2\pi f\tau} \tag{4.2}$$

where z(t) represents the analytic signal defined in (1). The importance of the use of this analytic signal is described in [11].

The properties of the WVD are well described in several comprehensive papers, such as [37] and [10]. In particular, the WVD exhibits a range of peaks along the curve describing the instantaneous frequency law of the signal, which allows frequency tracking of time-varying signals.

Inspection of Equation 4.2 indicates that the evaluation of the WVD is a non-causal operation. Thus, the value of the signal should be known for all time before either the Hilbert transform or the WVD can be evaluated. In many practical cases, the signals considered have either a finite duration or band limited spectrum and the problem is overcome by applying the WVD to a windowed version of the signal z(t) at time t_0 such that $z_m(t,t_0) = z(t)m(t-t_0)$, where $m(t)$ is a time-limited window function with duration T. The WVD of the windowed signal is given by a one-dimensional convolution in the frequency variable, f:

$$W_{z_m}(t,f) = W_z(t,f) \star_f W_m(t,f) \tag{4.3}$$

where $\star_f$ denotes frequency convolution.

Thus the effect of windowing is to smear the WVD representation in the frequency direction only; hence the frequency resolution can be increased by using a longer window without adversely affecting the time resolution. This is a significant improvement over the magnitude squared short-time Fourier transform (STFT) $S_{z_m}(t,f)$ where time resolution is degraded by the window as shown by the following relationship

$$|S_{z_m}(t,f)|^2 = W_z(t,f) \star_{tf} W_m(t,f) \tag{4.4}$$

where $\star_{tf}$ denotes convolution in time and frequency.

For notational convenience, we define K_z and k_m such that:

$$K_z(t,\tau) = z(t+\tau/2)z^*(t-\tau/2) \text{ and } k_m(\tau) = m(\tau/2)m^*(-\tau/2) \quad (4.5)$$

The WVD at any point in time is evaluated by shifting the signal z(t) so that the window is centered about $t = 0$. Therefore, the WVD computation reduces to the evaluation of Equation 4.6:

$$\hat{W}_{z_m}(t,f) = \int_{-T/2}^{+T/2} K_z(t,\tau)k_m(\tau)e^{-i2\pi f\tau}d\tau \quad (4.6)$$

where $\hat{W}_z$ is an estimator of W_z.

Equation (6) shows that the estimation of the WVD reduces to the computation of the FT of a signal kernel $K_z(t,\tau)$ as a function of τ with an analysis window $h(t) = k_m(\tau)$, which is a classical problem of signal analysis. An efficient implementation of (6) which uses fundamental symmetry properties of the WVD is described in [9].

4.6 The Autoregressive WVD (AWVD)

4.6.1 Description of the method

Equation (6) shows that WVD analysis may be thought of as a two-step process:

1. A sequence of kernels $K_z(t,\tau)$ is formed.
2. A Fourier Transform of $K_z(t,\tau)$ is taken, with respect to the variable τ, using the analysis window k_m to reduce spectral leakage.

It is easily shown that for the linear FM signal, $K_z(t,\tau)$ becomes a single complex sinusoid with frequency varying with the chirp parameter α determining the slope [7]. That is, if

$$z(t) = e^{i2\pi(f_c t+\alpha t^2/2)}, \quad (4.7)$$

then the WVD kernel becomes:

$$K_z(t,\tau) = e^{i2\pi(f_c+\alpha t)\tau} \quad (4.8)$$

Since a harmonic process can be generated by an autoregressive (AR) model, the real part of the kernel $K_z(t,\tau)$ at time t may be fitted by the output of an appropriate AR model; the AR coefficients will estimate the cross-section of the Wigner-Ville Distribution at time t. The real part is taken because the kernel has an Hermitian symmetry and therefore, no additional information is contained in the imaginary part. The analytic signal must however be used in the estimation of the kernel. The expression for the kernel is then:

$$K_z(t,\tau) = s(t+\tau/2)s^*(t-\tau/2) + \mathcal{H}[s(t+\tau/2)]\mathcal{H}[s^*(t-\tau/2)] \qquad (4.9)$$

where $\mathcal{H}[.]$ represents the Hilbert Transform. A variety of methods for super-resolution spectral estimation are available for use with this method, based on parametric modelling.

i) Autoregressive modelling

A method of spectral estimation which has been widely accepted in seismic [15] and speech processing [33] is the method of auto-regressive modelling. Makhoul [33] shows that the autoregressive modelling is equivalent to solving the normal equations associated with the linear prediction problem for a finite set of observed samples. For the optimum predictor to use only a finite number of samples of the past, the prediction error filter will have only zeros, and the spectrum being estimated must have only poles. It is the use of this structural information which permits the use of maximum entropy estimators when the process model is appropriate [33].

The traditional approach for implementing this method consists of

1. calculating the sample autocorrelation or autocovariance function
2. solving a linear least-squares problem to find the optimum predictor tap weights
3. Fourier transforming the prediction error filter weights.

This method assumes that the sampled kernel data under analysis can be expressed as a pth order AR model. Then the N point truncated kernel must satisfy the difference equation

$$k_z(n) = \epsilon(n) - \sum_{i=1}^{p} a(i)k_z(n-i) \qquad (4.10)$$

where N is odd and

$$\begin{aligned} k_z(n) &= z(t+L)z^*(t-L) \\ L &= n - \left(\frac{N-1}{2}\right), \end{aligned} \qquad (4.11)$$

$\epsilon(k)$ is a sample of the noise which is assumed to come from a white noise process; $a(i)$, $i = 1 \ldots p$ are the coefficients of the AR model. (Here, we have assumed a change of variable $n = \tau/2$ before discretizing, as reported in [11].)

Then the covariance AR parameter estimates, $\hat{a}$ and $\hat{\sigma}$, are given by the solution of

$$C\hat{\mathbf{a}} = \hat{\sigma}^2 \mathbf{e}_1 \tag{4.12}$$

where C is the covariance matrix formed by the elements

$$c(i,j) = \frac{1}{N-p} \sum_{s=p}^{N-1} k_z^*(s-i) k_z(s-j) \quad for\ i,j = 0 \ldots p \tag{4.13}$$

$$\mathbf{e}_1 = [1\ 0 \ldots\ 0\ ]^T$$

and

$$\hat{\mathbf{a}} = [1\ \hat{a}(1) \ldots\ \hat{a}(p)\ ]^T \ ; \ (p+1\ elements)$$

Equation 4.12 is conceptually solved as follows :

$$\begin{aligned} \mathbf{b} &= C^{-1}\mathbf{e}_1 \\ \mathbf{b} &= \frac{\hat{\mathbf{a}}}{\hat{\sigma}^2} \end{aligned} \tag{4.14}$$

Since $a(0) = 1$, $\hat{\sigma}^2 = \frac{1}{b(0)}$ and $\hat{\mathbf{a}} = \hat{\sigma}^2 \mathbf{b}$.

Unfortunately C^{-1} does not exist if C is singular. For this reason, the Moore-Penrose pseudoinverse, C^+, is used. This may be computed from the nonzero eigenvalues, λ_k, and corresponding eigenvectors, $\mathbf{v}_k$, of C as follows:

$$C^+ = \sum_{s=1}^{M} \frac{1}{\lambda_s} \mathbf{v}_s \mathbf{v}_s^H \tag{4.15}$$

where the eigenvalues are ranked in decreasing magnitude $\lambda_1 \geq \lambda_2 \ldots \lambda_p$ and there are m estimated principle components ($m \leq p$). The value m represents the effective rank of the matrix C.

ii) Singular Value Decomposition

As already stated, we chose not to use the standard normal equation solution to the prediction problem because of its known sensitivity to ill conditioning. The calculations shown in this paper were made using the SVD method proposed by Tufts & Kumaresan [68] and subsequently studied by Minami, Kawaka and Minami [39]. The use of the SVD permits the use of a regularized version of the Moore Penrose generalized inverse in which the regularization can be controlled by setting the small singular values to zero when they are within a threshold, typically a small multiple of machine precision - macheps (smallest number represented by a given computer which when added to 1 returns something greater than 1). This threshold is a function of the SNR, and should be adjusted accordingly.

The AWVD is implemented by using the singular value decomposition (SVD) of the kernel data matrix to compute a reduced rank principle eigenvector approximation to the inverse covariance matrix for use in the AR modelling process. Since C can be expressed as the product of the observation matrix, K, and its Hermitian transpose, the eigenvectors and eigenvalues of C can be determined from a singular value decomposition of K, i.e.

$$C = K^H K \tag{4.16}$$

where K is the matrix defined by the elements $k(i,j) = K_z(p+i-j)$, for $i = 0 \ldots\ N-1-p,\ j = 0 \ldots\ p$. Singular value decomposition of K gives

$$K = U \Sigma V^H \tag{4.17}$$

where U and V are unitary and Σ is a diagonal matrix with the elements, $\sigma(k,k)$, (called the singular values of K) ranked in decreasing order such that $\sigma(1,1) \geq \sigma(2,2) \ldots \geq \sigma(p,p)$. Then the eigenvectors of C correspond to the columns of V and the eigenvalues of C are given by the squares of the singular values of K.

iii) The Algorithm: Details of the Implementation

For a given value of t, eq (10) can be expressed as: $-K\mathbf{a} + \mathbf{n} = \mathbf{k}$, with the following notation:

$$\mathbf{a} = \begin{bmatrix} a_p(p) \\ \vdots \\ a_1(p) \end{bmatrix}$$

$$\mathbf{n} = \begin{bmatrix} n_p(p) \\ \vdots \\ n_1(p) \end{bmatrix}$$

$$\mathbf{k} = \begin{bmatrix} k_p(p) \\ \vdots \\ k_1(p) \end{bmatrix}$$

$$K = \begin{bmatrix} k(0) & k(1) & k(2) & \dots & k(p-1) \\ k(1) & k(2) & k(3) & \dots & k(p) \\ k(2) & k(3) & k(4) & \dots & k(p+1) \\ \vdots & \vdots & \vdots & \vdots & \vdots \\ k(N-p-1) & . & . & . & k(N-2) \end{bmatrix} \qquad (4.18)$$

K is seen to be a Hankel matrix where the elements of the matrix are the same along the reverse diagonal and whose first row is $\mathbf{k}_z^T$.

In the absence of noise, the estimate a can be given by:

$$\mathbf{a} = -K^{+}\mathbf{k}. \qquad (4.19)$$

where K^{+} is the generalized inverse of the matrix K obtained via the reduced rank SVD, i.e. is obtained using the inverse of singular values above the threshhold and the corresponding eigenvectors.

This solution corresponds to a least-squares fit of the data to a model using a finite number p of non-harmonically related sinusoids in noise, where p is the order of the autoregressive model. The parameter p is estimated by either the numerical rank of the data matrix K as computed via the SVD, or is estimated by information theoretic methods. The reader is asked to compare this result with the observation in [34] that the discrete Fourier Transform of a data sequence is equivalent (up to a normalization factor) to a least-squares fit with a model which is composed of a finite number (N, size of the DFT) of harmonically related sinusoids. It is through the inclusion of the knowledge that we are looking for p non harmonically non-related sinusoids instead of N harmonically related sinusoids that the improved (high-precision and accuracy) performance of this Autoregressive Wigner-Ville estimation results.

Selection of the rank, m, can be made by determining the number of the m largest singular values which represent primarily the signal components and are usually well separated from the p-m-1 small values which represent the noise. In practice, to represent the spectrum of a signal which consists of L sinusoids, we need a rank of at least $2L$. Slight overestimation of the rank usually does not affect significantly the resulting spectral estimate. The smallest singular values corresponding to the noise (assumed to be equal for white noise), determine the value of σ.

The window length should be chosen such that it is large enough to contain the majority of the signal, but bounded by its finite duration. The window used is at most the natural effective window $m(\tau)m(-\tau)$ of the kernel occuring in the WVD process [8].

The cross-section of the Wigner-Ville Distribution at time t is then estimated by:

$$W(t,f) = \frac{\hat{\sigma}^2}{|1 + \sum_{n=1}^{p} \hat{a}(n)e^{-i2\pi fn}|^2} \tag{4.20}$$

where $\hat{\sigma}^2$ and $\hat{a}(n)$ are determined from a kernel, k_z, centered about time t, where $t = n/f_s$.

4.6.2 Implementation of the AWVD

Conventional WVD analysis as defined in (2) is performed using a software package based on the algorithm described in [8]. In the new method, the estimated AWVD is calculated as shown in Figure 1 using a peak-finding algorithm to ensure accurate representation of the extremely sharp peaks. Two important features of the implementation are:

1. AWVD performance and speed is improved by performing the modelling on the real part of the kernel since the imaginary part contains redundant information. However the analytic signal must still be used to form the kernel, so that low frequency artifacts are not created [11].

2. The analytic signal must be interpolated before forming the kernel.

4.6.3 Results and discussion

i) Linear FM

The signals considered in this section are FM signals. The linear FM signal can be expressed as:

$$z(t) = e^{j2\pi(f_c t + \alpha t^2/2)} \tag{4.21}$$

for which the WVD kernel becomes:

$$k_z(t,\tau) = e^{i2\pi(f_c + \alpha t)\tau}. \tag{4.22}$$

It is clear that this kernel is composed of one single frequency at any particular time, and that a spectral analysis of the windowed kernel will be more effective than the spectral analysis of the windowed signal since the windowed signal will always contain more than one spectral component.

The test signal is a 512 sample linear FM signal sampled at 200 Hz and modulated from a lower frequency of 10 Hz to an upper frequency of 90Hz. In this implementation, the analytic signal is calculated in the frequency domain and is interpolated by zero padding before performing the inverse discrete time Fourier Transform [11]. The AWVD results are shown in Figure 2 and can be compared with the WVD result in Figure 3.

ii) Hyberbolic FM

The hyperbolic FM signal is expressed as:

$$z(t) = e^{\frac{i2\pi f_0 \log(|1+at|)}{a}} \tag{4.23}$$

where f_0 is the initial frequency of the sweep at $t = 0$ and f_c is the center frequency.

Its instantaneous frequency is:

$$f_i(t) = \frac{f_0}{1+at} \tag{4.24}$$

for which the WVD kernel becomes:

$$K_z(t,\tau) = e^{\frac{i2\pi f_0}{a} \log\left\{\frac{|1+a(t+\tau/2)|}{1+a(t-\tau/2)|}\right\}}$$

By dividing both numerator and denominator by $2(1+at)$, the expression for the log can be expanded as:

$$log\left|1+\frac{a\tau}{2(1+at)}\right| - log\left|1-\frac{a\tau}{2(1+at)}\right|$$

which can be simplified if we have the condition: $\frac{a\tau}{2(1+at)} \ll 1$.

In practice, the signals under analysis are causal and have a finite duration T. Therefore, the maximum value of the numerator in the above inequality is aT. From the expression for $f_i(t)$ given in Equation 4.24, we can also see that the minimum value of the denominator is 1. The previous inequality is therefore valid when $aT \ll 2$ is satisfied. From the expression for $f_i(t)$ in (24), when t equals T, then $f_i(t)$ equals $f_0 + B$. This yields the relation $aT = \frac{B}{(f_0+B)} \ll 2$. By introducing the signal center frequency f_c, this condition can be expressed as:

$$B \ll f_c \tag{4.25}$$

This condition corresponds to narrow band signals. We thus obtain

$$K_z(t,\tau) \simeq e^{i2\pi\left(\frac{f_0}{1+at}\right)} = e^{i2\pi f_i(t)\tau} \tag{4.26}$$

and the same comments as in the linear FM case apply.

results

The AWVD and WVD were applied to an increasing hyperbolic FM signal and the results are shown in figures 4 and 5. The peaks of the AWVD follow the frequency law more closely than the WVD does and the peak width is typically reduced by a factor of 5 to 10. Since the Covariance AR modelling is not subject to windowing distortion, the multiple sidelobes evident in the rectangularly windowed WVD do not appear in the AWVD. Improvement in the frequency law tracking of the AWVD over the conventional WVD is evident. The autoregressive modelling of the WVD kernel data is better than the AR modelling of the signal data because the bilinear form ensures that within a window, we have only one frequency (see equation at the beginning of this section), whereas with the signal data, we always have several frequencies present in the window.

iii) Multicomponent Signals

Although it is expected the the AWVD would perform badly for this class of signals, because of the cross-terms generated by the bilinear kernel, it is nevertheless interesting to present the following example. A multi-component signal formed from the sum of two chirp signals is analyzed by the AWVD and the WVD. and the results are shown in fig.6 and fig.7. This example shows the cross-product "ghost" peaks which alternate between positive and negative values in the WVD, but are always positive in the AWVD as well as being reduced. The artifacts found in the WVD analysis of multicomponent signals are reduced by the use of the AWVD because the components of the Kernel which give rise to the artifacts are not well modelled by the Autoregressive model which was chosen to model the components of the kernel corresponding to the "autoterms".

The robustness of the method to parameter selection and to additive noise has been tested [12]. For chirp signals analyzed in noise, the results are summarized as follows:

1. For signals with a large SNR, the result is a high accuracy estimate of the WVD, which is largely independent of the model order and rank.

2. The length of the window appears not to affect the resolution of the WVDs.

3. The rank should be approximately the same as (or greater by 1 than) the model order.

4. The higher the model order and rank the greater the resolution of the peaks, but if the model order becomes too large, spurious peaks may occur.

5. The model breaks down below about 5-10 db SNR. This is a reasonable result since the method for estimating the WVD peaks is similar to Prony's method for estimating sinusoids in noise – it can only be successful if the original signal is dominant, *i.e.* the SNR is large enough.

4.6.4 Applications of the AWVD

This section is intended to provide the reader with a few examples of applications of this method, which provided the reason for developing this AWVD estimator.

1. Seismic: As the echolocation search for petroleum resources becomes more difficult, the seismic industry must resolve geological structures with increasing accuracy. Since the bandwidth of the signals is limited by the propagation (attenuation) properties of the earth, modern spectral estimation techniques are playing an ever more important role in seismic prospecting. The high-accuracy AWVD described here provides one method for achieving better identification of fine strata; and this can provide enhanced interpretation of seismic sections, revealing more detail about the thin layers. The modern methods of seismic exploration are now based upon the use of linear FM signals as source instead of classical explosives. These FM signals are emitted by multiple large trucks with vibrating plates in contact with the surface of the ground. This signal fits the model used for the derivation of the AWVD estimator. Under the assumption of the convolution model (weak reflection), then the Wigner-Ville Distribution of the received signal is the convolution (in time) between the WVD of the emitted signal and the WVD of the sequence of reflectors. If the WVD of the sequence of reflectors was known, then using the well-known invertability of the WVD, the sequence of reflectors can be recovered up to a multiplicative constant. Thus the practical problem of Wigner-Ville analysis of the seismic problem is a classical inverse scattering problem in which it is desired to deconvolve the instrument response function, i.e. the WVD of the transmitted signal from the WVD corresponding to the sequence of reflectors. Fourier deconvolution procedures are not appropriate in this case, and high accuracy Wigner-Ville Analysis using the AWVD may be thought of as being equivalent to a numerically robust solution to the classical inverse scattering problem.

2. Sonar/radar problem: The sonar/radar echolocation problem is a generalization of the seismic problem described above in which the indi-

vidual reflectors may also be in motion so that they both provide a delay and a frequency shift of the signal. This problem is normally described in terms of the Ambiguity-function of the transmitted waveform which may be viewed as the output of a matched filter matched to the signal waveform when that waveform is delayed and frequency-shifted. It is well-known [37] that the WVD is equal to the 2DFT of the ambiguity function, and therefore the whole echolocation system can be described in terms of the WVD [74]. In the case where the scattered signals are modelled as Gaussian random variables, the WVD at the output of the receiver may be written as the 2D convolution of the WVD of the transmitted signal with the scattering function of the reflectors in range and doppler. This, too, is now a 2D scattering problem where the instrumentation function, ie; the WVD of the transmitted signal, approximates a line integral, eg. a projection of the scattering distribution when the transmitted signal is a linear FM, and thus the equivalent deconvolution problem is anhomorphic (there is different accuracy in range and doppler). Also, the deconvolution is enhanced by high-accuracy techniques which make the line integral approximation more closely valid, so that the deconvolution can be achieved by tomographic reconstruction techniques [74].

3. Fresnel region beamforming: The third example is an example of the passive localisation problem where multiple spherical waves are received by a uniformly spaced line-array of sensors such as geophones, hydrophones or electromagnetic antennas. It has been shown in the array processing literature that when the receiving array is in the far field, the spherical waves may be approximated by plane waves. Then the distribution of phase along the sensors will be linear and different ratios of phase shifts correspond to different angles of arrival; and thus the problem of estimating the direction of arrival is equivalent to the spectral analysis of the time signal corresponding to the spatial samples of the signal. This type of signal processing is equivalent to decomposing the incoming signal field into a frequency-wavenumber spectrum [28]. When the sources are in the Fresnel zone of the array, then the curvature of the spherical wavefront must be properly accounted for. Inspection shows that it is now a linear FM, which again corresponds to the signal for which the proposed AWVD method will be most effective [13][66]. Although WVD analysis allows for the simultaneous estimation of direction of arrival and range, a high-accuracy WVD is required to provide practical estimation of the range.

LIST OF FIGURES AND CAPTIONS

Figure 1: Time-Frequency Representation of a linear FM chirp signal obtained by using an SVD-based Autoregressive Wigner-Ville Distribution (AWVD).

Figure 2: Time-Frequency Representation of a linear FM chirp signal obtained by using the conventional Wigner-Ville Distribution (WVD).

Figure 3: Time-Frequency Representation of a hyperbolic FM chirp signal obtained by using an SVD based Autoregressive Wigner-Ville Distribution (AWVD).

Figure 4: Time-Frequency Representation of a hyperbolic FM chirp signal obtained by using the conventional Wigner-Ville Distribution (WVD).

Figure 5: Time-Frequency Representation of two cross-linear FM chirp signals obtained by using an SVD based Autoregressive Wigner-Ville Distribution.

Figure 6: Time-Frequency Representation of two cross-linear FM chirp signals obtained by using the conventional Wigner-Ville Distribution (WVD).

4.7 Radar Imaging

A new formulation of tomographic imaging for rotating targets is presented in terms of the Wigner-Ville distribution. This new formulation both simplifies the signal processing required for image generation and is applicable in both mono-static and bi-static situations. Also, the new formulation will allow the use of parametric modifications of the WVD, including the AWVD described above, to improve the resolution attainable in radar imaging.

The target is assumed to have multiple scatterer within each range-Doppler resolution cell so that the received signal will be a sample function of a Gaussian random variable. When the transmitted waveform is a linear frequency modulated chirp then the expected value of the output of a matched filter receiver is the two-dimensional convolution of the range-Doppler scattering function of the target with the auto-ambiguity surface of the transmitted waveform. Under the additional assumption that the transmitted signal has a Gaussian or rectangular envelope then the receiver can be simplified to a short-time Fourier transform receiver using only *a priori* information about the envelope and the chirp rate of the transmitter. This simplification eliminates the need for detailed phase information at the receiver and thus bi-static operation is feasible. Since most surfaces are rough at optical frequencies this form of signal processing should be particullarly useful with optical sensors such as coherent laser radars (LIDARs).

4.7.1 Introduction to Range-Doppler Imaging

Range-Doppler imaging (scatterer distribution mapping) is a general class of inverse problem which arises in remote sensing applications in either radar or sonar whenever the sensor is in relative motion with respect to the object or objects under study. In principle the sensors can radiate either electromagnetic (radar) or acoustic (sonar) signals. The radiated signals are usually in the form of a modulation on a carrier wave. The ratio of modulation bandwidth to carrier frequency is called the fractional bandwidth. We can distinguish between two cases, depending on the fractional bandwidth:

1. Small Fractional Bandwidth: The fractional bandwidth is usually small in the case of radar and thus motion induced Doppler can be treated as a single frequency shift of the entire modulation spectrum.

2. Large Fractional Bandwidth: The fractional bandwidth is often large in the case of sonar and thus motion induced Doppler should be treated as a change in time scale of the modulation signal or equivalently the Doppler shift caused by the motion is proportional to the frequency of the components of the transmitted spectrum.

For simplicity we will consider only the small fractional bandwidth case and will use radar terminology and examples. However, the signal processing concepts presented in this section can be extended to the large fractional bandwidth case for use in sonar signal processing.

The problem which will be the subject of this paper is the mapping of rigid objects which are in motion with respect to a stationary sensor. Although rigid objects are the primary targets of interest in most remote sensing applications non rigid scatter distributions such as a flight of birds or a school of fish may be mapped if the overall shape of the swarm does not change too much during the mapping operation. By specalizing to rigid objects, it is possible in many applications to incorporate *a priori* or in some cases *a posteriori* information about the target and thus convert the measured scatterer distribution to a range and cross-range image of the object. Two types of scattering, deterministic and stochastic, have been considered in the literature. This paper will assume the stochastic model of scattering which is usually valid when the target is rough with respect to the wavelength of the illuminations. Thus the inverse problem is to estimate the range-Doppler scattering function of the target which represents the average power reflected from a resolution cell of the target. If the target is essentially two-dimensional, (i.e. a rotating disk), when viewed along the line of sight (LOS) of the radar than the range-Doppler cells of the scattering function form a one-to-one mapping with range and cross-range cells of the object. If the target is essentially three-dimensional, (i.e. a rotating sphere), then the mapping is no longer one-to-one and in general there may be two or more range and cross-range cells with the same range and Doppler (e.g. the rotation axis of the sphere is perpendicular to the LOS of the radar).

Although the analysis of the tomographic imaging method in this paper is stated in terms of a monostatic radar, (i.e. transmitter and receiver either co-located or closely located), the analysis can be extended to a bi-static radar, (transmitter and receiver at different locations). In the bi-static case the transformation from range and Doppler to range and cross-range is more difficult due to the geometrical terms in the transformation [3]. However, the measurement of the range-Doppler scattering function is still simple when the analysis is performed using the Wigner-Ville distribution instead of the ambiguity surface, (i.e. magnitude squared of the ambiguity function). As will be subsequently shown, the estimation of the scattering function separates into a measurement of the Wigner-Ville distribution of the received signal made at the receiver and the utilization of a prori information about the transmitted signal.

Although the concept presented in this paper is based on a stochastic model of the reflection process, it still works as described even when the target is deterministic and not fluctuating. In this latter case the signal processing proposed in this paper may be simplified by reducing the number of different chirp signals used and thus may be useful at microwave frequencies where longer coherent integration times are required due to the reduced Doppler sensitivy of the lower carrier frequency, compared to the LIDAR case. Also, since the nominal resolution of the range-Doppler cell is proportional to the bandwidth (range resolution) and the duration of coherent integration (Doppler resolution) use at microwave frequencies may require stepped frequency chirp signals instead of continuous chirp signals to satisfy the bandwidth limitations of available components [71].

4.7.2 WVD Background for Echography

In order to understand the role of the Wigner-Ville distribution in radar/sonar signal processing, it is necessary to introduce a model of the received signal which is sufficiently realistic to be practical yet sufficiently simple so that it can be easily understood. Therefore, only scalar signals will be considered. This assumption is true for acoustic waves propagating in a liquid medium and is a simplifying assumption for electromagnetic waves.

The received signal will be modeled as the linear superposition of scaled versions of the transmitted signal delayed by the round trip time and modified by Doppler. Since the transmitted signal has only limited range and Doppler resolution then the received signal may be simplified by assuming that the scale factor is the product of a constant that depends on the range and a complex random variable which acounts for interference between unresolved target scatterers and propagation multipaths. A reasonable assumption, when the individual contributions are sufficiently numerous and approximately equal in magnitude with uniformly distributed phase, is that the random variable can be modeled as a zero mean Gaussian with variance $\sigma(t, f)$ which depends on the range and Doppler. This function is traditionally called the scattering function and this target model has been described by VanTrees [69].

The scattering function will be assumed to be uncorrelated in frequency due to orthogonality of the Doppler shifts for narrow band signals. However, different delays may be correlated depending on the characteristics of the target. In order to be able to include both deterministic as well as stochastic variation from pulse to pulse the random variable will be modeled with an exponential time correlation which can be generated by a recursive filter. Conventional processing only works well when the correlation is high, (i.e. the target scattering is essentially deterministic). Speckle noise is associated with single look SAR images.

4.7.3 Comparision of Radar Imaging Techniques

Four different analyses of the imaging of radar targets have been proposed in the literature.

In the first and simplest the received signal from a spatially distributed target is illuminated at successive frequencies as a function of angle. If the scattering is assumed to be deterministic then the *two-dimensional Fourier transform* of the received signal gives the range and cross-range components of the scattering distribution provided the target is in the far field of the illumination and the variation in angle is sufficiently small that a rectangular approximation of a polar grid is valid. This type of signal processing is often used in radar ranges where the angular orientation can be measured and the signal to noise ratio is sufficiently large that matched filter processing is not required.

In the second method *matched filtering* is used with either continuous or stepped frequency chirp signals. Succesive range compressed signals from a moving target are aligned in time in order to compensate for the variation in range between succesive measurements and then a discrete Fourier transform is computed on the complex matched filter outputs for successive measurements. This Fourier transform computes the Doppler frequency as a function of range and is related to the cross-range of a rigid object if the effective azimuth rotation rate relative to the line of sight (LOS) is known. This method is effective for deterministic signal returns, such as those from metallic aircraft illuminated by microwave frequencies. However, when the target is rough relative to the illuminating wavelength then the scattering is stochastic and this type of processing is not appropriate.

In the third method the *cross-ambiguity function* between the transmitted and the received signal is computed. For deterministic scattering distributions this integral is proportional to the two-dimensional Fourier transform of the product of a complex reflection coeficient with the ambiguity function of the transmitted waveform–but where the proportionality factor is complex and range dependant. This method of range-Doppler imaging has been analyzed by Feig and Grunbaum and leads to a new form of tomographic imaging [22].

In the fourth method, the *magnitude squared of the cross-ambiguity function* between the transmitted and the received signal is computed. For stochastic scatterers the expected value of the magnitude squared of the cross-ambiguity function is proportional to the two-dimensional convolution of the range- Doppler scattering function with the magnitude squared of the auto-ambiguity function of the transmitted signal. Thus for rotation relative to the LOS the effective Doppler of the target becomes coupled to the range of the target through the ambiguity function's magnitude squared response. The latter is sometimes referred to as the ambiguity surface. For narrowband signals of unit energy the ambiguity surface is everywhere positive and has unit volume. Thus it may be considered to be a probability density function which distributes the energy from the scatterer in both range and Doppler acording to the particular wave form used by the transmitter. This method of imaging is analagous to the medical method of time-of-flight positron emmision tomography (TOFPET) when finite duration chirp signals are used and is analogous to the medical method of computerized axial tomography (CAT) in the limit of infinite duration chirp signals. The analysis of this method of signal processing will be considered next.

4.7.4 Tomographic Radar Imaging using Ambiguity Surfaces

In 1984, Bernfeld proposed the use of a computerized axial tomography (CAT) analogy to process linear fm synthetic aperture (SAR) images [5]. Instead of transmitting repeatedly the same linear fm signal as the radar moves along its path, Bernfield suggested that chirps of different slopes be transmitted. For a stochastic Gaussian assumption of the scattering process the output of a matched filter receiver may be modeled as the two-dimensional convolution of the range-Doppler scattering function with the auto-ambiguity surface of the transmitted signal. In the limit of an infinite duration linear fm signal and no additive noise the output of the receiver may be considered to be equivalent to a sequence of line integrals through the scattering function where the orientation of the particular line integral relative to the scattering distribution is determined by the chirp rate of the transmitted signal. The ambiguity surface of an infinite duration linear fm signal may be considered to be a delta function with an orientation in range and Doppler determined by the chirp rate of the signal. Therefore, the convolution of this ambiguity function with the scattering distribution may be considered to be a sequence of tomographic projections of the scattering distribution.

In most sonar and radar applications, infinite duration signals are not practical.Therefore, the tomographic signal processing technique proposed by Bernfield should be modified to accomodate finite duration signals. For a finite duration linear fm signal the ambiguity surface is only an approximation to a line integral. In 1985 Snyder and Whitehouse proposed a signal processing modification based on the medical method of time-of-flight positron emission tomography (TOFPET) [58].

When a linear fm signal with a Gaussian envelope is used as the transmitted signal in an echo ranging system then the auto-ambiguity surface of the transmitted signal is an asymmetric two-dimensional Gaussian function. The inverse problem which results when a number of different orientations of the two- dimensional Gaussian are convolved with an unknown but bounded two-dimensional distribution has been studied in the context of reconstructing medical radio nucleide activity distributions [57]. One of the techniques of solving this inverse problem is called the "confidence weighting" algorithm and is a generalization of the CAT technique of back projection and superposition followed by two-dimensional filtering.

A phenomological description of confidence weighting can be easily constructed to motivate the confidence weighting tomographic algorithm. In the conventional CAT algorithm there is no information in the projection or line integral about the location of the contributions to the line integral. Therefore, the value of the line integral is uniformly distributed along the path of the line integral, (i.e. back projection). In the case of TOFPET or the proposed tomographic radar imaging algorithm the pointspread function of the tomograph or the ambiguity surface provides *a priori* information about the contributions to the receiver output. Since the normalized ambiguity surface has unit volume and is positive, it may be interpreted as a probability distribution and thus confidence weighting is simply a method analogous to back projection for incorporating the *a priori* knowledge.

4.7.5 WVD Interpretation of Imaging

It is well known that the Wigner-Ville distribution (WVD) and the ambiguity function of Sussman are related by a two-dimensional Fourier transform [67]. Thus it is not surprising that radar or sonar detection and estimation can be reinterpreted in terms of the WVD. However, it is not without some difficulty as will be presented in the following section. But the difficulties are minor compared to the insight provided by reformulating the signal processing equations. In particular, such an interpretation of the signal processing for bi-static operation of the radar or sonar is easier to understand, since the effect of receiver mismatch may be predicted and in the case of linear frequency modulated (LFM) signals simplified signal processing either in terms of the short time Fourier transform (STFT) or STRETCH processing is possible.

The WVD has many interesting properties [37]. However, only a few will be needed for our purposes. These are:

- Realness
- The shift properties
- Moyal's formula
- The marginal properties
- Relationship to the ambiguity function
- Relationship to the ambiguity surface

Unfortunately, there are some additional properties which would be desirable in a time-frequency distribution, but which are attainable only when other desirable prorerties are sacraficed. These are:

- Linearity; Must give up realness or resolution.
- Positivity; Must give up resolution,the marginal properties, and the general applicability of Moyal's formula.

Tomographic signal reconstruction techniques restores linearity and as well as positivity. High resolution Wigner-Ville analysis shows promise in restoring the resolution [12]. However, Moyal's formula is no longer applicable, and therefore there is a small loss of detectability which is not usually important for imaging applications [23].

The tomographic processing algorithm using the Wigner-Ville distribution can be simply deduced from the corresponding tomographic algorithm using the ambiguity surface. The expected value of the cross-ambiguity surface of the matched filter receiver is the two-dimensional convolution of the scattering function with the auto-ambiguity function of the transmitted waveform [69]. Using the relationship between the ambiguity surface and the Wigner-Ville distribution the cross-ambiguity surface is rewritten as the two-dimensional convolution of the Wigner-Ville distribution of the received signal with the Wigner-Ville distribution of the transmitted signal. In a simillar manner the auto-ambiguity surface of the transmitted signal is the two-dimensional convolution of the Wigner-Ville distribution with itself. If the two-dimensional Fourier transform (i.e. ambiguity function) of the WVD of the transmitted signal is positive then the an equivalent formulation is that the expected value of the Wigner-Ville distribution of the received signal is the two-dimensional convolution of the target scattering function with the Wigner- Ville distribution of the transmitted signal.

LASER detection and ranging (LIDAR) technology has significant resolution advantages over conventional microwave radar techniques. The technology has received considerable attention as a means of imaging spaceborne objects using inverse synthetic aperture radar (ISAR) processing techniques using range resolved tomographic reconstruction. These ISAR techniques, however, become ineffective when the object is only viewed for a small fraction of a rotation, a common occurrence in strategic defense applications. Furthermore, the formidable signal processing requirements inherent in a ISAR imaging application require innovative techniques to minimize the size, weight, and power of the signal processor, particularly for spaceborne systems. Recently, Whitehouse and Boashash have suggested that a LIDAR imaging processor which uses tomographic reconstruction techniques to form an image may overcome the limited viewing angle problem associated with range resolved tomographic ISAR processing [75]. In this new technique, the chirp rate of the LIDAR transmitted pulse is varied during the illumination period. For these signals, the Wigner-Ville distribution, which simultaneously measures the signal time-delay and Doppler shift, is known to be an approximation to a knife-edge whose orientation in the time-delay, Doppler plane is a function of the chirp rate. Thus the output of a Wigner-Ville receiver which is the convolution of this rotated function with the received LIDAR signal and corresponds to calculating a projection of the object's scattering distribution. Time of flight positron emmision tomography (TOFPET) reconstruction techniques can then be employed to reconstruct an image of the object by confidence weighting, (i.e. convolving the received Wigner-Ville distribution with the Wigner-Ville distribution of the illuminating signal). and filtering the superposition of the confidence weighted signals.

4.8 Conclusions

The direction finding algorithms we have considered impose a substantial computational load for their real-time implementation. Most of this computational load consists of matrix operations:

- matrix-vector multiplication
- linear equation solution
- orthogonal triangularization
- triangular backsolve
- Hermitian symmetric eigensystem solution
- singular value decomposition
- Hermitian symmetric generalized eigensystem solution

These matrix operations also comprise a substantial part of the computational load for several useful methods for data compression, spectral analysis, and pattern recognition.

4.8.1 Research Issues in Parameter Estimation

Two of most promising parameter estimation techniques for further study for application to direction finding are Cadzow's method and the Maximum Likelihood method. MUSIC and MVDR should be viewed as baseline algorithms for comparison with the newer methods. Both Cadzow's eigenvector fitting method and the Maximum Likelihood Method permit high resolution direction finding in the presence of fully coherent multipath, and both are applicable to very general array geometries. However, both require the solution of a computationally expensive nonlinear least squares problem. At present, it is not only difficult to specify a preferred method for solving this nonlinear least squares problem, but even to specify the number of arithmetic operations needed for an acceptable solution. In order to make Cadzow's method and MLM feasable for real-time implementation and examine their robustness, further study is needed for:

- Convergence of Nonlinear Least Squares Algorithms
- Parallelization of Nonlinear Least Squares Algorithms
- NLS solution accuracy needed for Cadzow's Method and MLM

- Effect of element gain, phase, and position errors on Cadzow's Method and MLM

For the special case of radar imaging via estimation of the target scattering function, time-frequency analysis via the application of signal subspace techniques to the Wigner-Ville distribution leads to problems analogous to spectral estimation [76]. For such problems, the spectral analysis of the WVD's bilinear kernel is analagous to direction finding using a uniformly spaced line array, and ESPRIT and its variants are strong candidate methods. These techniques will require parallel implementations for:

- The Nonsymmetric Generalized Eigenvalue Problem
- Total Least Squares
- The Orthogonal Procrustes Problem

4.8.2 Parallel Computation

Surveys of parallel algorithms and architectures for many of the matrix computations discussed in this paper are available [62], [64]. Although nonlinear least squares algorithms use the preceding operations, the resulting iterative algorithms are not as well understood as the classical matrix operations. Additional research is also needed in the area of parallel algorithms for the nonsymmetric eigenvalue problem, since current techniques using parallel Jacobi-like methods for computing the Schur decomposition can have convergence problems for non-normal matrices. [20]

Acknowledgement

One of the authors (HJW) wishes to thank Prof. Donald Snyder of Washington University for many stimulating discussions of tomographic techniques.

One of the authors (BB) wishes to acknowledge the support of the Australian Research Council.

One of the authors (JMS) wishes to acknowledge the support of the NOSC Internal Research Program.

Bibliography

[1] H. C. Andrews and C. L. Patterson, "Outer Product Expansions and Their Uses in Digital Image Processing", *American Math. Monthly,* Vol.82, No.1, January 1975, pp. 1-13

[2] H. C. Andrews and C. L. Patterson, "Singular Value Decomposition (SVD) Image Coding", *IEEE Trans. on Communications,* April 1976, pp. 425-432

[3] O. Arikan and D.C. Munson, Jr., "A tomographic formulation of bistatic synthetic aperture radar," in W.A Porter and S.C. Kak, Editors, *Proceedings of the 1988 international conference on advances in communication & control systems*, Dept. of Electrical and Computer Engineering, Louisiana State University, Baton Rouge, LA 70803, October 19-21, 1988, Baton Rouge, LA, pp. 418-431

[4] William P. Ballance, "The Singular Value Decomposition: An Introduction", *Conference Record of the eighteenth Asilomar Conference on Circuits, Systems & Computers*, held at Pacific Grove, California, Nov.5-7, 1984, IEEE Catalog No.85CH22200-4, pp. 148-157

[5] M. Bernfeld, Chirp Doppler radar, *Proc. IEEE Lett.*, Vol 72, No.4, April 1984, pp. 540-541

[6] G. S. Bienvenu, "Influence of the Spatial Coherence of the Background Noise on High Resolution Passive Methods", *ICASSP 79 Record*, pp. 306-309

[7] Boashash, B. and H.J. Whitehouse, "Seismic Applications of the Wigner-Ville Distribution", *Proceedings of IEEE International Symposium on Circuits and Systems*, Vol. 1 of 3, pp. 34-37, San Jose, USA, May 1986

[8] B. Boashash and H.J. Whitehouse, "High Resolution Wigner-Ville Analysis", *Onzième Colloque GRETSI*, Nice, 1-5 June, 1987, pp. 205-208

[9] Boashash, B. and P. Black, "An efficient real time implementation of the Wigner-Ville Distribution", *IEEE transactions on Acoustics, Speech and Signal Processing*, Vol. 35, pp. 1611-1618, November, 1987

[10] Boashash, B. "Time-Frequency Signal Analysis and Synthesis - the choice of a Method and its Application", *Proc. of SPIE Conference, Advanced Algorithms and Architectures*, San Diego, Vol. 975, pp. 164-185, Aug. 1988

[11] Boashash, B., "Note on the Use of the Wigner Distribution for Time Frequency Signal Analysis", *IEEE Trans. on ASSP*, Vol. 36, pp. 1518-1521, September, 1988

[12] Boashash, B. and H. J. Whitehouse, "High Accuracy Time-Frequency Analysis of FM Signals using an Autoregressive Wigner-Ville Distribution", submitted to the *IEEE Transactions on Acoustics, Speech, and Signal Processing*

[13] B. R. Breed and T. E. Posch, "A Range and Azimuth Estimator Based on Forming the Spatial Wigner Distribution", *Proc. ICASSP '84*, paper 41B9

[14] Y. Bressler and A. Macovski, "Exact Maximum Likelihood Parameter Estimation of Superimposed Exponential Signals in Noise", *IEEE Trans. on Acoustics, Speech, and Signal Processing*, Vol. ASSP-34, No.5, Oct. 1986, pp. 1081-1089

[15] Burg, J.P., "Maximum Entropy Spectral Analysis", *Proc. 37th Meet. Soc. of Exploration Geophysicists*, Oklahoma City, OK, 1967

[16] J. A. Cadzow, "A High Resolution Direction-of-Arrival Algorithm for Narrow-Band Coherent and Incoherent Sources", *IEEE Trans. on Acoustics, Speech, and Signal Processing*, Vol.36, No.7, July, 1988, pp. 965-979

[17] J. Capon, R. Greenfield, and R. J. Kolker, "Multidimensional Maximum-Likelihood Processing of a Large Aperture Seismic Array", *Proc. IEEE*, Vol.55, p.192- 211, Feb. 1967

[18] J. Capon, "High Resolution Frequency-Wavenumber Spectrum Analysis", *Proc. IEEE*, Vol. 57, pp. 1408-1418, Aug.1969

[19] Cohen, L., "Time-Frequency Distributions: A Review", *Proc. IEEE*, July 1989

[20] P. J. Eberlein, "On the Schur Decomposition of a Matrix for Parallel Computation", *IEEE Trans. on Computers*, Vol. C-36, No.2, Feb. 1987, pp. 167-174

[21] M. Feder and E. Weinstein, "Optimal Multiple Source Location Estimation via the EM Algorithm", *ICASSP 85 Proceedings*, Vol.4, paper 45.6, pp. 1762-1765

[22] E. Feig and F.A Grünbaum, "Tomographic methods in range-Doppler radar," *Inverse Problems*, Vol. 2, pp. 185-195, 1986

[23] P. Flandrin, "A Time-Frequency Formulation of Optimum Detection", *IEEE Trans. on Acoustics, Speech, and Signal Processing*, Vol.36, No.9, September 1988, pp. 1377-1384

[24] G. H. Golub and C. F. Van Loan, *Matrix Computations*, Johns Hopkins University Press, Baltimore, 1983, third printing, with corrections, 1985

[25] I. J. Good, "Some Applications of the Singular Decomposition of a Matrix", *Technometrics*, Vol.11, No.4, November 1969, pp. 823-831

[26] S. Haykin (Ed.), *Nonlinear Methods of Spectral Analysis*, Springer-Verlag, 1979

[27] D. H. Johnson and S. R. De Graaf, "Improving the Resolution of Bearing in Passive Sonar Arrays by Eigenvalue Analysis", *IEEE Trans. on Acoustics, Speech, and Signal Processing*, Vol. ASSP-30, No.4, pp. 638-647, August 1982

[28] Kailath, T.(Ed.), *Modern Signal Processing*, Hemisphere Publishing Corporation, 1985

[29] F. T. Luk, "A Parallel Method for Computing the Generalized Singular Value Decomposition", *Journal of Parallel and Distributed Computing*, Vol2, 1985, pp. 250-260

[30] M. T. Ma, *Theory and Applications of Antenna Arrays*, John Wiley & Sons, 1974

[31] R. N. McDonough, "Application of the Maximum-Likelihood Method and Maximum Entropy Method to Array Processing", Chapter 6 of *Nonlinear Methods of Spectral Analysis*, edited by S. Haykin, Springer-Verlag, 1979

[32] J. G. McWhirter and T. J. Shepherd, "Efficient minimum variance distortionless response processing using a systolic array", *Proc. SPIE*, Vol.975, paper 975-39, pp. 385-392

[33] Makhoul, J., "Linear Prediction: A Tutorial Review", *Proc. IEEE*, Vol. 63, No.4, pp. 561-580, April, 1975

[34] Marple, S.L., *Digital Spectral Analysis - with applications*, Prentice Hall Inc, 1985

[35] S. L. Marple, *Digital Spectral Analysis*, Prentice-Hall, 1987

[36] G. E. Martin, " Signal subspace processing of experimental radio data", *SPIE Proceedings* Vol.975, paper 975-11, pp. 101-107, 1988

[37] W. Mecklenbräuker, "A Tutorial on Non-Parametric Bilinear Time-Frequency Signal Representations", in *Les Houches, Session XLV, 12 Août-6 Septembre 1985, Traitment du Signal / Signal Processing*, edited by J. L. Lacoume, T. S. Durrani, and R. Stora, Elsevier Science Publishers, B. V., 1987, pp. 277-336

[38] M. I. Miller and D. R. Fuhrmann, "An EM Algorithm for Direction-of-Arrival Estimation for Narrowband Signals in Noise", *Proc. SPIE*, Vol.826, pp. 101-103 (1987)

[39] Minami, K., Kawata, S. and Minami, S., "Superresolution of Fourier Transform Spectra by Autoregressive Model Fitting with Singular Value Decomposition", *Appl. Optics*, Vol. 24, no. 2, pp. 162-167, Jan. 1985

[40] R. A. Monzingo and T. W. Miller, *Introduction to Adaptive Arrays*, John Wiley & Sons, 1980

[41] J. C. Nash, *Compact Numerical Methods for Computers*, John Wiley & Sons, New York, 1979

[42] U. Nickel, "Angle Estimation with Adaptive Arrays and Relation to Superresolution", *IEE Proc.*, Vol.134, Part II, pp. 77-82 (1987)

[43] N. L. Owsley, "Sonar Array Processing", Chapter 3 of *Array Signal Processing*, edited by S. Haykin, Prentice-Hall, 1985

[44] C. C. Paige, "Computing the Generalized Singular Value Decomposition", *SIAM J. Sci. Stat. Comput.*, Vol.7, No.4, October 1986, pp. 1126-1146

[45] A. Paulraj, R. Roy, and T. Kailath, "Estimation of Signal Parameters via Rotational Invariance Techniques - ESPRIT", *Conference Record of the Nineteenth Asilomar Conference on Circuits, Systems, and Computers*, Nov.6-8, 1985, pp. 83-89

[46] A. Paulraj, R. Roy, and T. Kailath, " A subspace rotation approach to estimating directions of arrival", *Proc. ICASSP 86*, April 7-11, 1986, Vol.4, paper 47.2, pp. 2495-2498

[47] A. Paulraj, R. Roy, and T. Kailath, "A subspace rotation approach to signal parameter estimation", *Proc. IEEE*, July 1986, pp. 1044-1045

[48] V. F. Pisarenko, "The Retrieval of Harmonics from a Covariance Function", *Geophys. J. R. astr. Soc.*, Vol.33, pp. 347-366, (1973)

[49] S. S. Reddi, "On a Spatial Smoothing Technique for Multiple Source Location", *IEEE Trans. on Acoustics, Speech, and Signal Processing*, Vol. ASSP-35, No.5, May, 1987, p.709

[50] R. Roy, A. Paulraj, and T. Kailath, "Comparative Performance of ESPRIT and MUSIC for Direction-of-Arrival Estimation", ICASSP 87, IEEE Order No. CH2396-0/87, Volume 4, paper 54.12, pp. 2344-2347

[51] R. Roy and T. Kailath, "Invariance Techniques and high-resolution null steering", *Advanced Algorithms and Architectures for Signal Processing III*, SPIE Vol. 975, (1988), paper 975-35, pp. 358-367

[52] R. O. Schmidt " Multiple Emitter Location and Signal Parameter Estimation", *Proceedings of the RADC Spectrum Estimation Workshop*, Rome Air Development Center, pp. 243-257, Oct.3-5, 1979

[53] R. O. Schmidt, *A Signal Subspace Approach to Multiple Emitter Location and Spectral Estimation,* Ph.D. Dissertation, Dept. of Electrical Engineering, Stanford University, November, 1981

[54] R. Schreiber and P. J. Kuckes, "Systolic Linear Algebra Machines for Digital Signal Processing", in *VLSI and Modern Signal Processing*, edite by S. Y. Kung, H. J. Whitehouse, and T. Kailath, Prentice Hall, 1985, pp. 389-405

[55] T. J. Shan, M. Wax, and T. Kailath, "On Spatial Smoothing for Direction of Arrival Estimation of Coherent Signals", *IEEE Trans. on Acoustics, Speech, and Signal Processing*, Vol.ASSP-33, No.4, Aug. 1985, pp. 806-811

[56] S. D. Silverstein, "Robust Spectral Estimation: Autocorrelation Based Minimum Free Energy Method", *Twenty-Second Asilomar Conference on Signals, Systems & Computers*, Oct. 31-Nov. 2, 1988

[57] D. L. Snyder, L. J. Thomas, Jr., and M. M. Ter-Possian, "A Mathematical Model for Positron Emission Tomography Systems Having Time-of-Flight Measurements", *IEEE Trans. on Nuclear Science*, Vol. NS-28, pp. 3575-3583, June 1981

[58] D.L. Snyder, H.J. Whitehouse, J.T. Wohschlaeger, R.C. Lewis, "A New Approach to RADAR/SONAR Imaging", *Proceedings of the SPIE 30th Annual Technical Symposium*, San Diego, CA, August 1986

[59] J. M. Speiser, H. J. Whitehouse, and N. J. Berg, " Signal Processing Architectures using Convolutional Technology", *Proc. SPIE* Vol.154, pp. 66-80, 1978

[60] Speiser, J., Bromley, K., and Whitehouse, H.J., "Signal Processing Applications for Real-Time Matrix Operations", in *Advanced Computer Concepts*, LJI Conference Proceedings, La Jolla Institute, Editor J.C. Solinsky, pp. 35-46, 1981

[61] J. M. Speiser, "Progress in Eigenvector Beamforming", *SPIE Proceedings* Vol.564, paper 564-01, 1985

[62] J. M. Speiser, "Signal Processing Computational Needs", *Proc. SPIE*, Vol.696, pp. 2-6 (1986)

[63] J. M. Speiser, "Some Observations Concerning the ESPRIT Direction Finding Method", *Proc. SPIE*, Vol.826, pp. 178-185, (1987)

[64] J. M. Speiser, "An Overview of Matrix-Based Signal Processing", *Conference Record of the Twenty-First Asilomar Conference on Circuits, Systems, and Computers*, held at Pacific Grove, CA, Nov.2-4, 1987, IEEE Computer Society Order No. 760, pp. 284-289

[65] B. D. Steinberg, *Principles of Aperture and Array System Design*, John Wiley & Sons, 1976

[66] A. L. Swindlehurst and T. Kailath, "Near-field source parameter estimation using a spatial Wigner distribution approach", *Advanced Algorithms and Architectures for Signal Processing III*, SPIE Vol. 975, (1988), paper 975-09, pp. 86-92

[67] H.H Szu, "Two-dimensional optical processing of one-dimensional acoustic data", *Optical Engineering* Vol. 21, 1982, pp. 804-813

[68] Tufts, D.W. and Kumaresan, R., "Singular value decomposition and improved estimation using linear prediction", *IEEE Trans ASSP*, vol.30, no.4, Aug 1982

[69] H.L. Van Trees, *Detection, Estimation, and Modulation Theory*, Part III, Radar-Sonar Signal Processing and Gaussian Signals in Noise, John Wiley and Sons, New York, 1971

[70] R. A. Wagstaff and A. Baggeroer (editors), *High Resolution Spatial Processing in Underwater Acoustics*, Naval Ocean Research and Development Activity, NSTL

[71] D.R. Wehner, *High Resolution Radar*, Artech House, 1987

[72] A. J. Weiss, A. S. Willsky, and B. C. Levy, "Maximum Likelihood Array Processing for the Estimation of Superimposed Signals", em Conference Record of the Twenty-First Asilomar Conference on Signals, Systems, and Computers, held at Pacific Grove, CA, Nov.2-4, 1987, Vol.2, pp. 845-848

[73] H. J. Whitehouse and B. Boashash, "Signal Processing Applications of Wigner-Ville Analysis", *SPIE Vol.696 Advanced Algorithms and Architectures for Signal Processing (1986)*, pp. 156-162

[74] Whitehouse, H. J., "Interpretation of radar imaging in terms of the Wigner-Ville distribution", presented at *Advanced Algorithms and Architectures in Signal Processing*, SPIE International Symposium, San Diego, Calif, August 15-17, 1988

[75] H. J. Whitehouse and B. Boashash, "Delay-Doppler Radar/Sonar Imaging", presented at EUSIPCO 88, Grenoble, France, 5-9 Sept, 1988

[76] H. J. Whitehouse and L. B. White, "Resolution Enchancement in Time-Frequency Signal Processing Using Tomographic Imaging", to appear in *ICASSP 89*, May 23-26, 1989, Glascow, Scotland

[77] I. Ziskind and M. Wax, "Maximum Likelihood Localization of Multiple Sources by Alternating Projection", *IEEE Trans. on Acoustics, Speech, and Signal Processing*, Vol. ASSP-36, No.10, Oct. 1988, pp. 1553-1560

[78] M. D. Zoltowski, "Novel Techniques for Estimation of Array Signal Parameters Based on Matrix Pencils, Subspace Rotations, and Total Least Squares", *ICASSP 88 Conference Proceedings*, Vol.V, pp. 2861-2864

[79] M. D. Zoltowski, "Generalized Minimum Norm and Total Least Squares with Applications to Array Signal Processing", *Advanced Algorithms and Architectures for Signal Processing III*, SPIE Vol. 975, (1988), paper 975-08, pp. 78-85

[80] M. D. Zoltowski and D. Stavrinides, "Sensor Array Signal Processing Via a Procrustes Rotations Based Eigenanalysis of the ESPRIT Data Pencil", *IEEE Transactions on Acoustics, Speech, and Signal Processing*, Vol. 37, No.6, June 1989, pp. 832-861

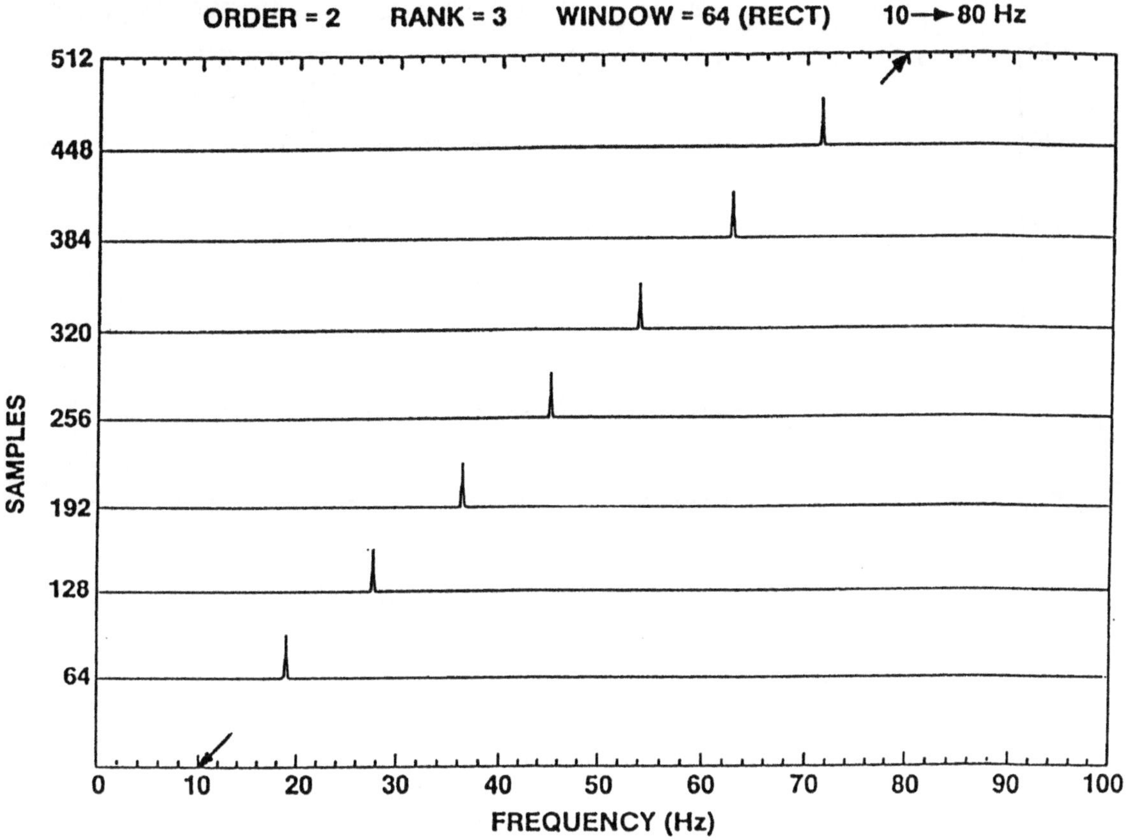

Figure 1: Time-Frequency Representation of a linear FM chirp signal obtained by using an SVD-based Autoregressive Wigner-Ville Distribution (AWVD).

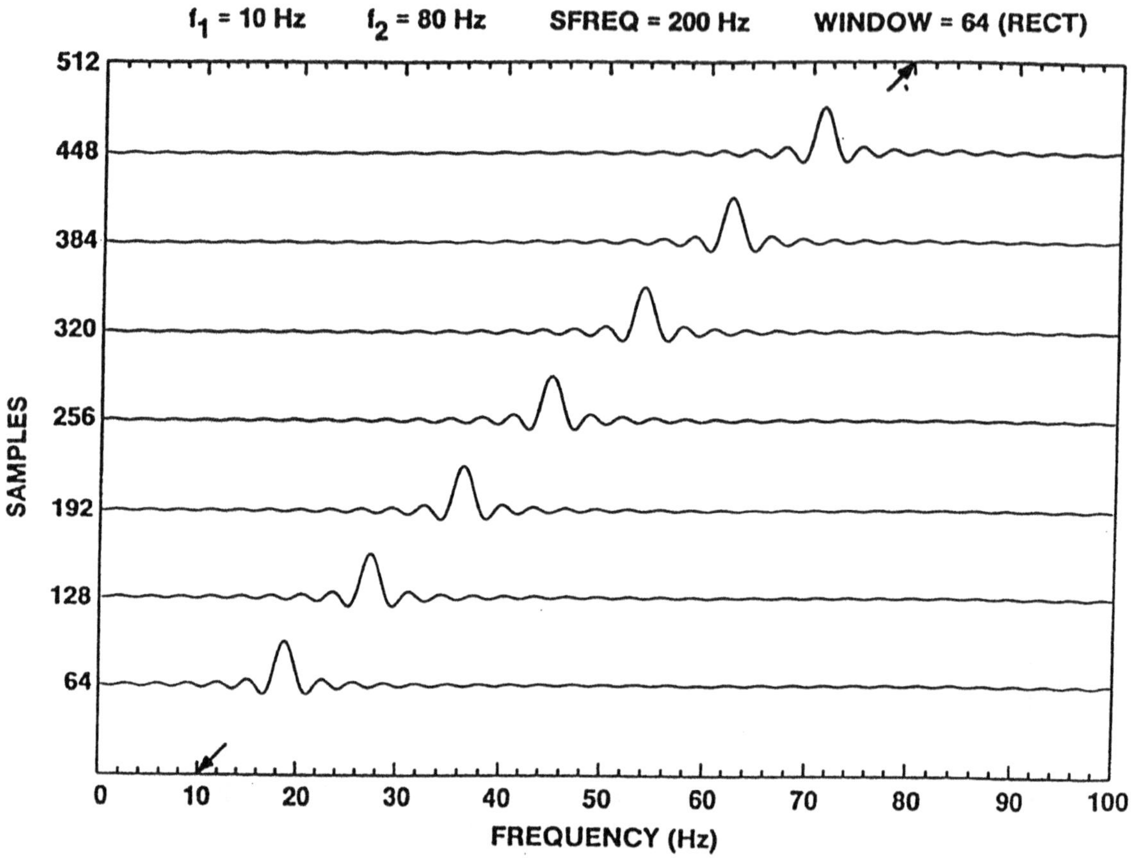

Figure 2: Time-Frequency Representation of a linear FM chirp signal obtained by using the conventional Wigner-Ville Distribution (WVD).

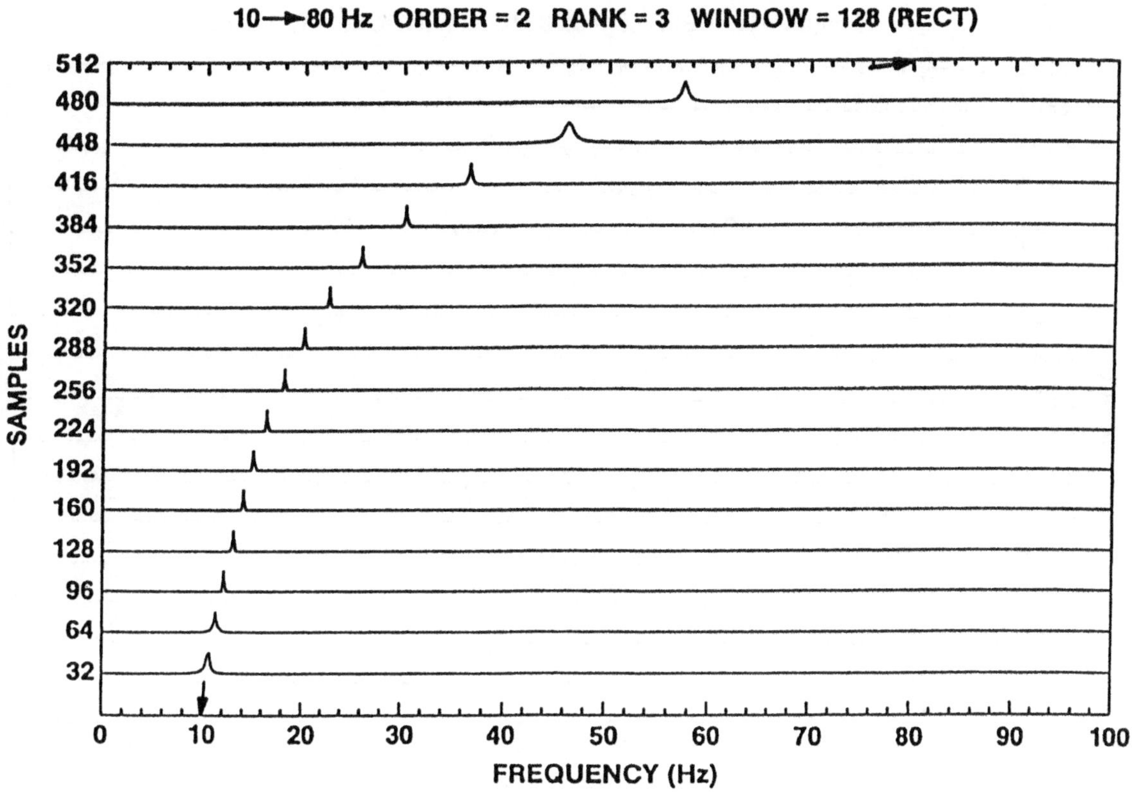

Figure 3: Time-Frequency Representation of a hyperbolic FM chirp signal obtained by using an SVD based Autoregressive Wigner-Ville Distribution (AWVD).

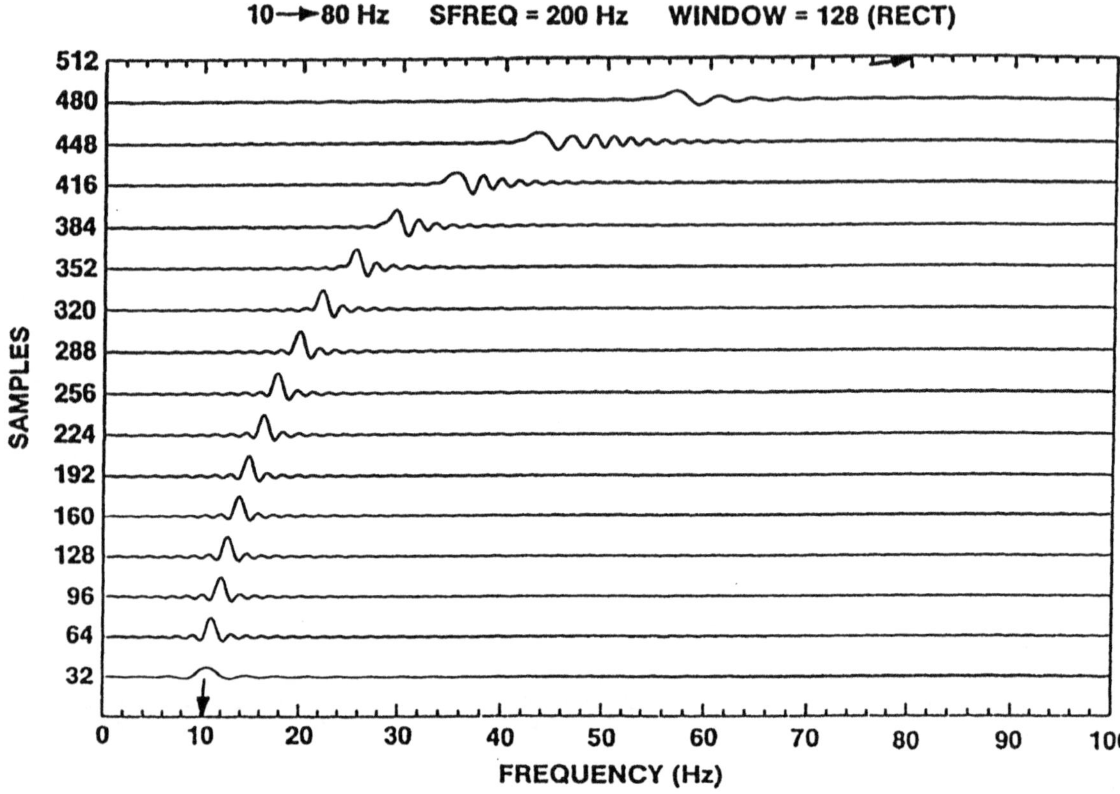

Figure 4: Time-Frequency Representation of a hyperbolic FM chirp signal obtained by using the conventional Wigner-Ville Distribution (WVD).

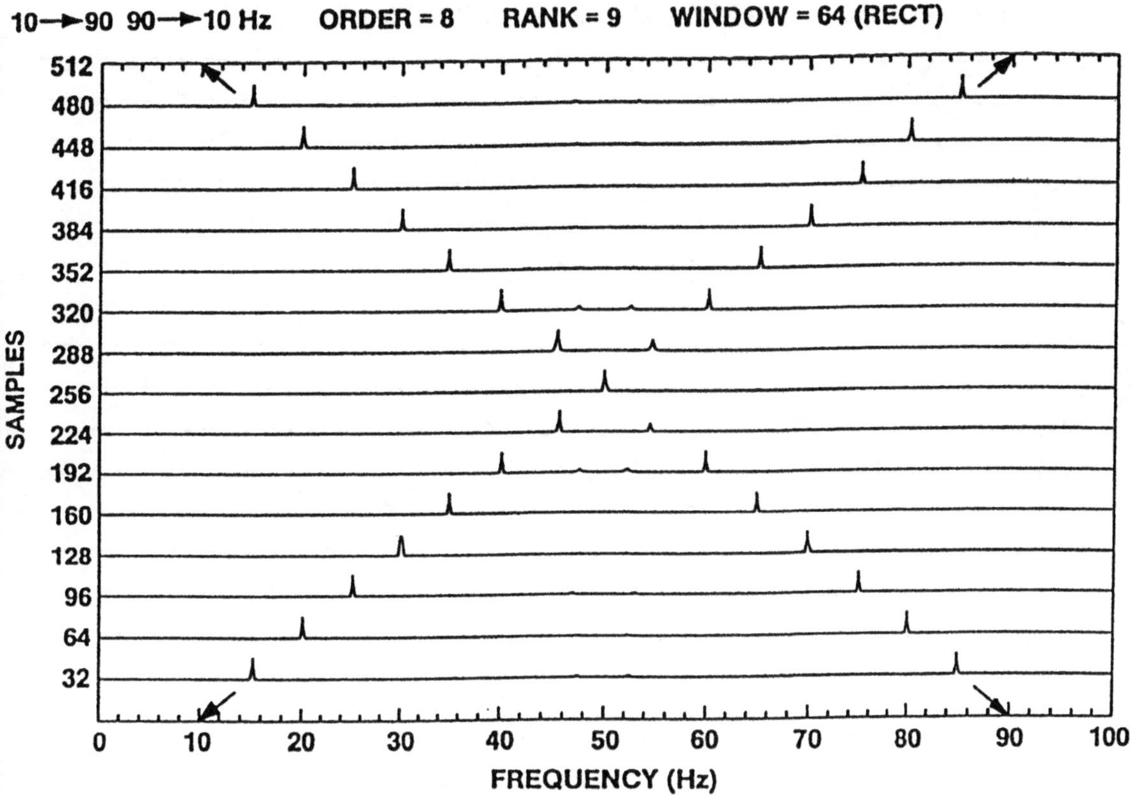

Figure 5: Time-Frequency Representation of two cross-linear FM chirp signals obtained by using an SVD based Autoregressive Wigner-Ville Distribution.

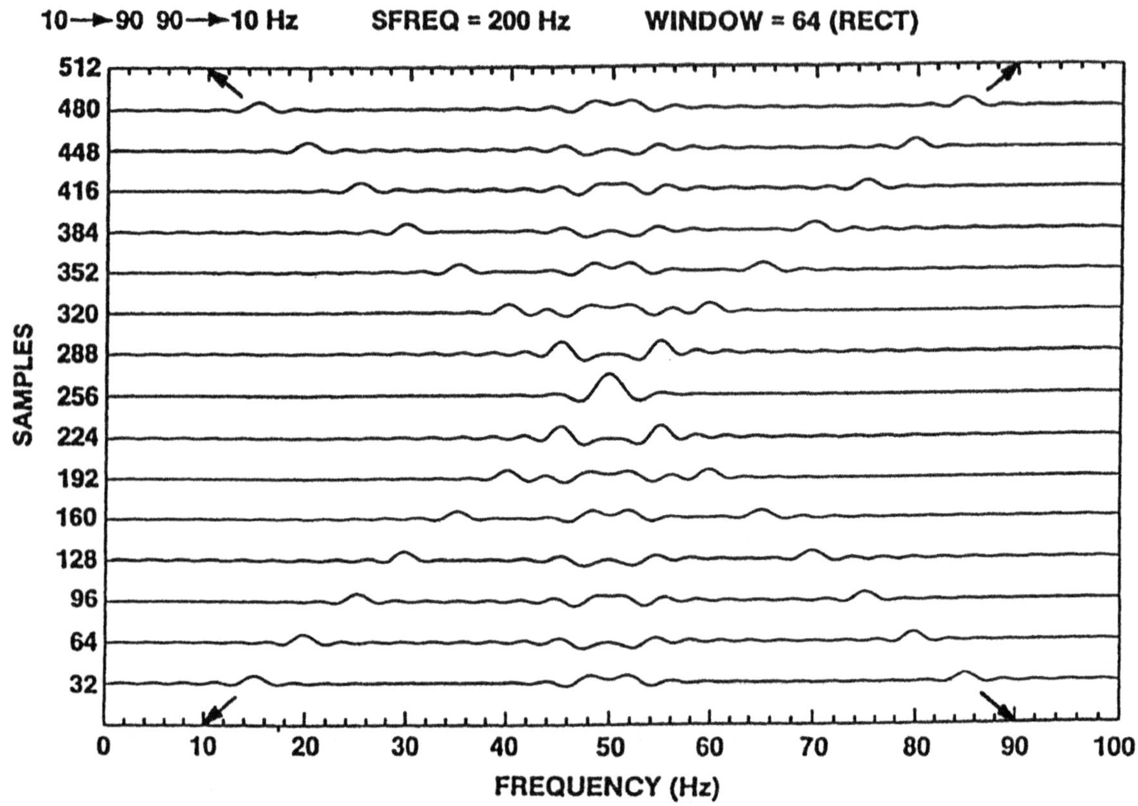

Figure 6: Time-Frequency Representation of two cross-linear FM chirp signals obtained by using the conventional Wigner-Ville Distribution (WVD).

CHAPTER 5

PERFORMANCE LIMITS OF HIGH-RESOLUTION SYSTEMS

A. BHUYAN
Tel Aviv University (Israel)

H. MESSER
Tel Aviv University (Israel)

P.M. SCHULTHEISS
Yale University (USA)

1. Introduction

This paper deals with the ability of a sensor array to resolve and/or locate closely spaced sources of random signals. In speaking of resolution it is necessary to distinguish between two somewhat different problems often addressed under that heading. The first [1]-[6] is basically a decision problem: Given a set of data, can one decide with reasonable confidence whether the number of sources is one or two (or more)? The second [7]-[10] is primarily an estimation problem: Given a set of data and the prior information that there are two sources, can one make meaningful statements about their separation? In this second setting one might say that the sources are not resolvable if the typical error in estimated separation exceeds the actual separation. The present paper takes the second point of view. It calculates instrumentation-independent lower bounds on the rms error of source separation and explores its dependence on actual separation, array geometry, knowledge concerning nuisance parameters such as signal power levels, and uncertainties in array geometry or propagation conditions. By carrying average source bearing as an additional parameter it also addresses the localization problem, because average bearing and angular separation are related to separate bearings by a simple linear transformation.

The basic geometry of the problem under consideration is shown in Fig. 1.

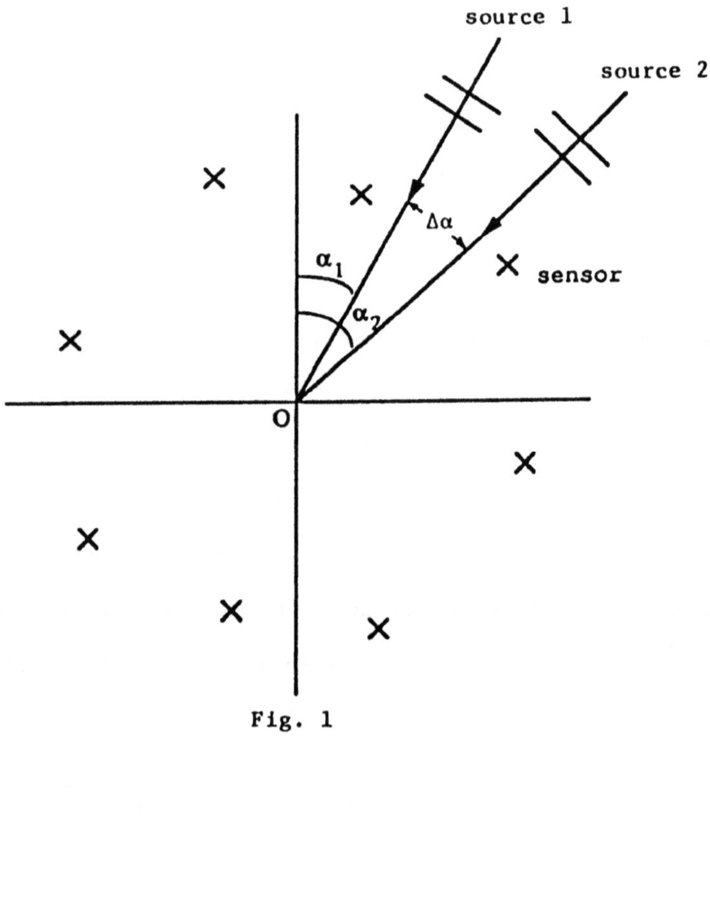

Fig. 1

The two sources radiate zero mean Gaussian noise spectrally flat over a band W. The observation time T is sufficiently large so that TW >> 1. We will work with a "narrowband" assumption, postulating a bandwidth W small enough so that the signal phases observed at each sensor do not change significantly with frequency. The sources will be located in the far field, so that their location is specified by the bearing pair (α_1,α_2) or alternatively by $(\alpha_{av},\Delta\alpha)$, the combination of average bearing and angular separation. The two source signals are statistically independent of each other and of all noise encountered by the sensors. The sensor noise is Gaussian, zero mean, spectrally flat over the signal band and statistically independent from sensor to sensor.

The primary tool of analysis will be the Cramer Rao Lower Bound (CRLB). It provides an absolute lower bound on the mean square error of any unbiased estimator and, under very general conditions, it is at least asymptotically (large data sets or high SNR) reachable by use of a maximum likelihood estimator [11]. This paper will not be concerned with establishing thresholds of SNR or observation time beyond which the CRLB is indeed reachable.

2. Basic Theory

The waveshape $y_i(t)$ received at the ith sensor is

$$y_i(t) = s_1(t-\tau_i) + s_2(t-\sigma_i) + n_i(t) \tag{1}$$

$s_j(t)$ is the signal radiated by source j and $n_i(t)$ is the noise component of the ith sensor output. The $n_i(t)$ and $s_j(t)$ are independent, zero mean, stationary and Gaussian random processes. τ_i and σ_i are the travel times of the two signals to sensor i. We assume that their values depend only on source - array geometry (straight-line propagation).

As stated in Section 1, we will assume that signal and noise spectra are flat over a band W centered at frequency ω_o (rad/sec) and that the observation time T satisfies TW >> 1. In that case the data are most conveniently represented by their Fourier coefficients

$$Y_{ki} = \int_{-\frac{T}{2}}^{\frac{T}{2}} y_i(t)e^{-j\omega_k t}dt, \qquad \omega_k = \frac{2\pi k}{T} \tag{2}$$

The Y_{ki} are then essentially uncorrelated for different indices k.[11] Since the Y_{ki} are also Gaussian, they can therefore be considered statistically independent.

If $\underline{Y}_k$ is the vector whose components are the Y_{ki} then from (1)

$$\underline{Y}_k = \sqrt{M}\ s_{1k}\,\underline{a}_{1k} + \sqrt{M}\ s_{2k}\,\underline{a}_{2k} + \underline{n}_k \tag{3}$$

s_{1k} and s_{2k} are the Fourier coefficients of the two signals and the components of $\underline{n}_k$ are the noise Fourier coefficients. M is the number of sensors.

$$\underline{a}_{1k} = \frac{1}{\sqrt{M}}\,(e^{j\omega_k\tau_1}\ \ldots.\ e^{j\omega_k\tau_M})^T \tag{4}$$

and

$$\underline{a}_{2k} = \frac{1}{\sqrt{M}}\,(e^{j\omega_k\sigma_1}\ \ldots.\ e^{j\omega_k\sigma_M})^T \tag{5}$$

are the normalized "steering vectors" of the two sources. It follows that the covariance matrix of $\underline{Y}_k$ is

$$K_k = MS_1(\omega_k)\underline{a}_{1k}\underline{a}_{1k}^H + MS_2(\omega_k)\underline{a}_{2k}\underline{a}_{2k}^H + N(\omega_k)I \tag{6}$$

$S_1(\omega)$, $S_2(\omega)$ are the signal spectra and $N(\omega)$ is the noise spectrum. $(\;)^H$ denotes the Hermitian transpose of the vector or matrix in parentheses. The complete data vector is the concatenation of the $\underline{Y}_k$, k = 1, 2.. N; its covariance matrix is block diagonal with the blocks K_k of (6) along the principal diagonal.

The parameters of interest (bearings and differential bearings) are specified by the delays τ_i and σ_i. The Cramer-Rao bound for any parameter vector $\underline{\phi}$ is given by the appropriate diagonal element of the inverse Fisher Information Matrix F^{-1}. The elements of F are

$$F_{ij} = -E\{\frac{\partial^2}{\partial\phi_i\partial\phi_i} \ln p(\underline{Y}/\underline{\phi})\} \tag{7}$$

$p(\underline{Y}/\underline{\phi})$ is the conditional probability density of the complete data vector $\underline{Y}$ given the parameters of interest. For the Gaussian case under consideration here one finds [12]

$$F_{ij} = \sum_{k=1}^{N} Tr\{K_k^{-1} \frac{\partial K_k}{\partial\phi_i} K_k^{-1} \frac{\partial K_k}{\partial\phi_j}\} \tag{8}$$

Tr { } stands for the trace of the matrix in brackets.

In the problem at hand, the steering vectors are related directly to the delays τ_i and σ_i, which themselves are functions of $\underline{\phi}$. It is therefore convenient to use the delays (or differences in delays) as intermediate parameters. In general, if $\underline{\phi} = \underline{g}(\underline{\theta})$, the Fisher Information for $\underline{\phi}$ is related to the Fisher Information for $\underline{\theta}$ by

$$F_{\underline{\phi}} = G \, F_{\underline{\theta}} \, G^T \tag{9}$$

where
$$G_{ij} = \frac{\partial \theta_i}{\partial \phi_j} \tag{10}$$

If the θ_i are delays and the ϕ_j are bearings, the G_{ij} depend only on source-array geometry.

As indicated before, we shall assume that the signal is "narrow-band" in the sense that the steering vectors are frequency invariant. If, in addition, all spectra are flat over a band of width W rad/sec and vanish outside of that band, Eq. (8) becomes

$$F_{ij} = \frac{TW}{2\pi} \, \mathrm{Tr} \, \{K_k^{-1} \frac{\partial K_k}{\partial \phi_i} K_k^{-1} \frac{\partial K_k}{\partial \phi_j}\} \tag{11}$$

Thus one is concerned only with the single frequency covariance matrix K_k.

3. Fisher Information for Time Delays

If the signal and noise spectral levels within the postulated narrow band are S_1, S_2 and N, one can use the spectral representation of the (rank 2) signal covariance matrix to rewrite (6) in the form

$$K_k = \lambda_1 \underline{V}_{1n} \underline{V}_{1n}^H + \lambda_2 \underline{V}_{2n} \underline{V}_{2n}^H + NI \tag{12}$$

The eigenvalues λ_1 and λ_2 are

$$\lambda_1 = \frac{M}{2}[(S_1+S_2) + \sqrt{(S_1-S_2)^2 + 4S_1S_2\xi^2}\,] \tag{13}$$

$$\lambda_2 = \frac{M}{2}[(S_1+S_2) - \sqrt{(S_1-S_2)^2 + 4S_1S_2\xi^2}\,] \tag{14}$$

where

$$\xi = \frac{1}{M}\,\underline{a}_1^H\underline{a}_2 \tag{15}$$

Thus $|\xi|^2$ is the well-known beam pattern, measuring the output power of a conventional beamformer steered in a direction corresponding to $\underline{a}_1$ when the signal comes from a direction specified by $\underline{a}_2$. ξ is, in general, a complex quantity. However, it is evident from (15) that ξ becomes real when the sensors of the receiving array maintain pairwise symmetry about the origin. In order to minimize algebraic complexity we impose this mild regularity condition on array geometry. As a further matter of analytical convenience (but without imposing additional physical constraints) we choose the origin at the center of gravity of the receiving array.

Unnormalized versions of the normalized eigenvectors $\underline{v}_{1n}$ and $\underline{v}_{2n}$ appearing in (12) are

$$\underline{v}_1 = \underline{a}_1 + \underline{a}_2\,\frac{1}{2S_1\xi}\,[\sqrt{(S_1-S_2)^2+4S_1S_2\xi^2} - (S_1-S_2)] \tag{16}$$

$$\underline{v}_2 = a_1 - \underline{a}_2\,\frac{1}{2S_1\xi}\,[\sqrt{(S_1-S_2)^2+4S_1S_2\xi^2} + (S_1-S_2)] \tag{17}$$

Eqs. (13), (14), (16) and (17) simplify dramatically when the power levels of the two sources are the same, i.e. $S_1 = S_2 \equiv S$. Then

$$\lambda_1 = MS(1+\xi) \quad , \quad \lambda_2 = MS(1-\xi) \tag{18}$$

$$\underline{v}_1 = \underline{a}_1 + \underline{a}_2 \quad , \quad \underline{v}_2 = \underline{a}_1 - \underline{a}_2 \tag{19}$$

From the definition of the steering vectors [Eqs. (4) and (5)]

$$\underline{v}_{1i} = \frac{2}{\sqrt{M}} \exp(-j\omega t_i) \cos \frac{\omega\Delta_i}{2} \tag{20}$$

$$\underline{v}_{2i} = -\frac{2j}{\sqrt{M}} \exp(-j\omega t_i) \sin \frac{\omega\Delta_i}{2} \tag{21}$$

where

$$t_i = \frac{\tau_i+\sigma_i}{2} \quad , \quad \Delta_i = \tau_i-\sigma_i \tag{22}$$

Thus t_i is the average signal travel time to sensor i and Δ_i the differential signal travel time. If $\underline{v}_o$ is the normalized steering vector associated with the mean location of the two sources

$$\underline{v}_{oi} = \frac{1}{\sqrt{M}} \exp(-j\omega t_i) \tag{23}$$

Defining diagonal matrices Λ_1 and Λ_2 whose diagonal elements are $\cos(\omega\,\Delta_i/2)$ and $\sin(\omega\,\Delta_i/2)$ respectively we can rewrite $\underline{v}_1$ and $\underline{v}_2$ [using (20) and (21)]

$$\underline{v}_1 = 2\,\Lambda_1\,\underline{v}_o \quad , \quad \underline{v}_2 = -2j\,\Lambda_2\,\underline{v}_o \tag{24}$$

Finally, observing that $||v_1||^2 = 2(1+\xi)$, $||v_2||^2 = 2(1-\xi)$, Eq. (12) becomes

$$K_k = 2MS\{(\Lambda_1 \underline{v}_o)(\Lambda_1 \underline{v}_o)^H + (\Lambda_2 \underline{v}_o)(\Lambda_2 \underline{v}_o^H)\} + NI \tag{25}$$

We note that $\underline{v}_o$ depends only on the average travel times (hence average source location) while Λ_1, Λ_2 depend only on differential travel time (hence source separation). For the equal power case, therefore, average location and separation are the natural parameter choices. The basic reason for this, is of course, that the eigenvectors of the signal covariance matrix are proportional to the average and the difference of the two steering vectors [Eq. (19)]. With unequal source powers, (16) and (17) suggest the use of weighted sums and differences. Because of the more immediate physical insights to be gained in the equal power situation, we work with that case for much of our discussion.

Using (25) in (11) the computation of the Fisher Information for the vectors of average delays $\underline{t}$ and differential delays $\underline{\Delta}$ is straightforward. One finds

$$F_{\underline{t},\underline{\Delta}} = \frac{TW}{2\pi} \begin{bmatrix} J_t & J_{t\Delta} \\ J_{t\Delta}^T & J_\Delta \end{bmatrix} \tag{26}$$

J_t, J_Δ and $J_{t\Delta}$ are the following MxM matrices

$$J_t = 4\omega^2 \{ \frac{2(\frac{S}{N})^2}{1+M\frac{S}{N}(1+\xi)} [\frac{M}{2}(1+\xi)\Lambda_1^2 - \Lambda_1^2 \underline{1}\,\underline{1}^T \Lambda_1^2]$$

$$+ \frac{2(\frac{S}{N})^2}{1+M\frac{S}{N}(1-\xi)} [\frac{M}{2}(1-\xi)\Lambda_2^2 - \Lambda_2^2 \underline{1}\,\underline{1}^T \Lambda_2^2]$$

$$- \frac{4(\frac{S}{N})^2 (1+M\frac{S}{N})}{[1+M\frac{S}{N}(1+\xi)][1+M\frac{S}{N}(1-\xi)]} \Lambda_1 \Lambda_2 \underline{1}\,\underline{1}^T \Lambda_2 \Lambda_1 \tag{27}$$

$$J_\Delta = \omega^2 \{ \frac{2(\frac{S}{N})^2}{1+M\frac{S}{N}(1+\xi)} [\frac{M}{2}(1+\xi)\Lambda_2{}^2 + \frac{1-M\frac{S}{N}(1+\xi)}{1+M\frac{S}{N}(1+\xi)} \Lambda_1\Lambda_2 \underline{1}\,\underline{1}^T \Lambda_2\Lambda_1]$$

$$+ \frac{2(\frac{S}{N})^2}{1+M\frac{S}{N}(1-\xi)} [\frac{M}{2}(1-\xi)\Lambda_1{}^2 + \frac{1-M\frac{S}{N}(1-\xi)}{1+M\frac{S}{N}(1-\xi)} \Lambda_1\Lambda_2 \underline{1}\,\underline{1}^T \Lambda_2\Lambda_1]$$

$$- \frac{2(\frac{S}{N})^2}{[1+M\frac{S}{N}(1+\xi)][1+M\frac{S}{N}(1-\xi)]} (\Lambda_1{}^2 \underline{1}\,\underline{1}^T\Lambda_2{}^2 - \Lambda_2{}^2 \underline{1}\,\underline{1}^T\Lambda_1{}^2)$$

$$- \frac{4(\frac{S}{N})^2}{[1+M\frac{S}{N}(1+\xi)][1+M\frac{S}{N}(1-\xi)]} [\frac{M}{2}(1+\xi)\Lambda_2{}^2\underline{1}\,\underline{1}^T\Lambda_2{}^2 - \frac{M}{2}(1-\xi)\Lambda_1{}^2\underline{1}\,\underline{1}^T\Lambda_1{}^2] \}$$

(28)

$$J_{t\Delta} = 0 \qquad (29)$$

The symbol $\underline{1}$ denotes an M vector all of whose elements are one.

Inspite of their algebraic complexity Eqs. (27)-(29) allow several immediate insights:

1) The fact that $J_{t\Delta} = 0$ indicates that the estimation errors of $\underline{t}$ and $\underline{\Delta}$ are uncoupled. Knowledge of one does not affect the ability to estimate the other. This is in contrast with the delays $\underline{\tau}$ and $\underline{\sigma}$ whose estimation errors become closely coupled when the source separation is small. It is important to keep in mind that the result $J_{t\Delta} = 0$ is dependent on the previously specified choice of the origin as the center of gravity of the array.

2) J_t and J_Δ depend on Δ but not on t.

3) When the source separation tends to zero, $\Delta_i \to 0$ so that $\Lambda_2 \to [0]$ while $\Lambda_1 \to I$ and $\xi \to 1$. In that limit

$$J_t = 4\omega^2 \frac{2M(\frac{S}{N})^2}{1+2M\frac{S}{N}} (I-\underline{1}\,\underline{1}^T) \tag{30}$$

$$J_\Delta = 0 \tag{31}$$

The former is simply the Fisher Information for a single source of double strength. The latter says that source separation becomes inherently unmeasurable when that separation becomes arbitrarily small.

It is physically obvious, though not immediately apparent from (27) and (28), that J_t and J_Δ are singular: If the radiated waveshapes are unknown, it is impossible to estimate the delays of each signal to each sensor and hence to estimate the τ_i and σ_i. What is measurable is the differences $\tau_i-\tau_j$ and $\sigma_i-\sigma_j$ and hence the bearings or differential bearings of the two sources.

4. Transformation to Location Parameters

If $\underline{t}$ and $\underline{\Delta}$ are convenient parameters to specify delays, the natural choice for location parameters is average bearing α_{av} and differential bearing $\Delta\alpha$. The Fisher Information for this pair of parameters is

$$F_{\alpha_{av},\Delta\alpha} = R\, F_{\underline{t},\underline{\Delta}}\, R^T \tag{32}$$

where

$$R = \begin{bmatrix} \frac{\partial t_1}{\partial \alpha_{av}} & \cdots\cdots & \frac{\partial t_M}{\partial \alpha_{av}} & \frac{\partial \Delta_1}{\partial \alpha_{av}} & \cdots\cdots & \frac{\partial \Delta_M}{\partial \alpha_{av}} \\ \frac{\partial t_1}{\partial \Delta\alpha} & \cdots\cdots & \frac{\partial t_M}{\partial \Delta\alpha} & \frac{\partial \Delta_1}{\partial \Delta\alpha} & \cdots\cdots & \frac{\partial \Delta_M}{\partial \Delta\alpha} \end{bmatrix} \tag{33}$$

The elements of R are determined entirely by source-array geometry. In compact notation

$$R = -\frac{1}{c} \begin{bmatrix} (\cos \frac{\Delta\alpha}{2})\, \underline{u}^T & (2 \sin \frac{\Delta\alpha}{2}) \underline{w}^T \\ (\frac{1}{2} \sin \frac{\Delta\alpha}{2}) \underline{w}^T & (\cos \frac{\Delta\alpha}{2}) \underline{u}^T \end{bmatrix} \tag{34}$$

If the coordinates of the ith sensor are (x_i, y_i) the components of $\underline{u}$ and $\underline{w}$ are

$$u_i = -x_i \cos \alpha_{av} + y_i \sin \alpha_{av} \tag{35}$$

$$w_i = x_i \sin \alpha_{av} + y_i \cos \alpha_{av} \tag{36}$$

Thus $\underline{u}$ and $\underline{w}$ dependent only on average bearing. It is apparent that R becomes near-diagonal as $\Delta\alpha \to 0$. For future reference we note also that

$$\frac{\partial \underline{u}}{\partial \alpha_{av}} = \underline{w} \quad , \quad \frac{\partial \underline{w}}{\partial \alpha_{av}} = -\underline{u} \tag{37}$$

The general expression for $F_{\alpha_{av}, \Delta\alpha}$ is therefore

$$F_{\alpha_{av},\Delta\alpha} = \frac{TW\cos^2\frac{\Delta\alpha}{2}}{2\pi c^2}\left[\begin{array}{c|c} \underline{u}^T J_t \underline{u} + (4\tan^2\frac{\Delta\alpha}{2})\underline{w}^T J_\Delta \underline{w} & 2\tan\frac{\Delta\alpha}{2}(\frac{1}{4}\underline{u}^T J_t \underline{w} + \underline{u}^T J_\Delta \underline{w}) \\ \hline 2\tan\frac{\Delta\alpha}{2}(\frac{1}{4}\underline{u}^T J_t \underline{w} + \underline{u}^T J_\Delta \underline{w}) & \underline{u}^T J_\Delta \underline{u} + (\frac{1}{4}\tan^2\frac{\Delta\alpha}{2})\underline{w}^T J_t \underline{w} \end{array}\right] \tag{38}$$

The vectors $\underline{u}$ and $\underline{w}$ have a simple physical interpretation evident from Fig. 2. The component u_i of $\underline{u}$ is the distance from the ith sensor to the average line of sight. The component w_i of $\underline{w}$ is the displacement of the ith sensor from the origin along the average line of sight. Thus for a line array broadside to the average line of sight $\underline{w} = 0$ and (38) becomes strictly diagonal.

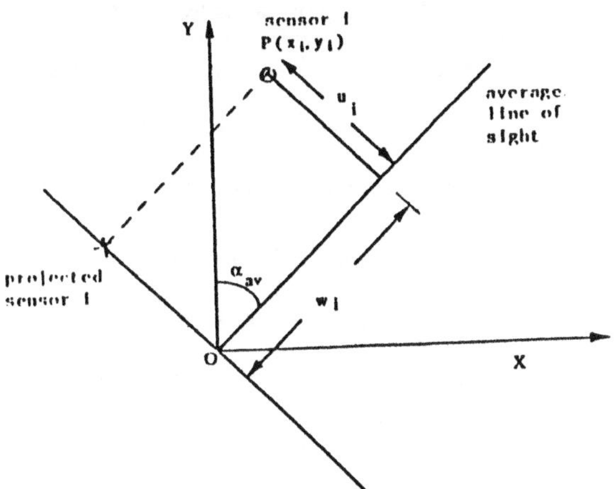

Fig. 2

In that case

$$F_{\alpha_{av},\Delta_\alpha} = \frac{TW}{2\pi c^2}\cos^2\frac{\Delta\alpha}{2}\begin{bmatrix} \underline{u}^T J_t \underline{u} & 0 \\ 0 & \underline{u}^T J_\Delta \underline{u} \end{bmatrix} \tag{39}$$

Precisely the same form is obtained for a two dimensional array in the limit as $\Delta\alpha \to 0$. The limiting situation is the case of a single stronger source with bearing α_{av}. For the single source case we have therefore proved: For bearing estimation purposes two arrays are equivalent if their sensor displacements from the line of sight are equal. In particular, one can obtain a broadside line array equivalent to any two dimensional array by projecting the sensors of the latter onto a straight line through the origin at right angles to the line of sight (see Fig. 2).

When there are two closely spaced sources one can attribute the small incremental terms in (38) to the fact that the projected arrays for the two sources are slightly different. Fig. 3 shows the relevant geometry. If one

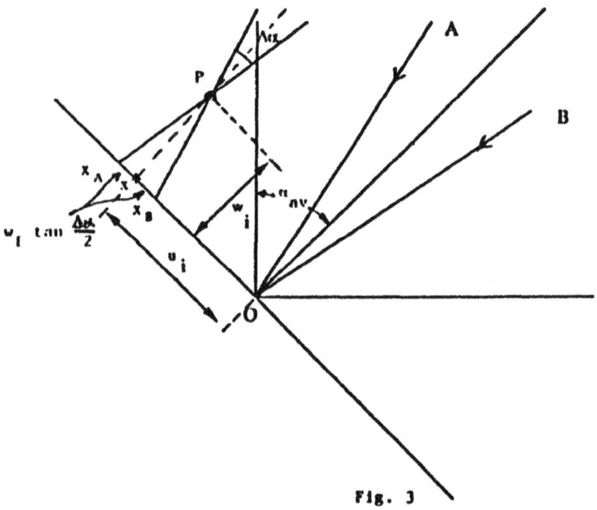

Fig. 3

chooses the line array at right angles to the average line of sight, the projections appropriate for the two sources differ by $w_i \tan \frac{\Delta\alpha}{2}$ from the average projection. If these displacements are small compared with distances between sensors in the average projected array, one would expect the errors arising from use of the average projected array to be small. More quantitatively, if $u_i >> w_i \tan(\Delta\alpha/2)$, all i, the diagonal terms of (38) are close to $\underline{u}^T J_{\underline{t}} \underline{u}$ and $\underline{u}^T J_{\underline{\Delta}} \underline{u}$ respectively [we shall find that $\underline{u}^T J_{\underline{\Delta}} \underline{u} \leq \underline{u}^T J_{\underline{t}} \underline{u}$] and the off diagonal terms are small compared with the diagonal terms.[1] Thus (39) remains a valid approximation. The only difficulty with this argument is that $\underline{u}^T J_{\underline{\Delta}} \underline{u} << \underline{u}^T J_{\underline{t}} \underline{u}$ for extremely small $\Delta\alpha$ [as suggested by (30) and (31)]. Hence the inequality

[1] The inequality $u_i >> w_i \tan(\Delta\alpha/2)$, all i, is clearly stronger than necessary. The quadratic form $\underline{u}^T J_{\underline{t}} \underline{u}$ is dominated by the largest u_i's, hence it is primarily for these that the inequality must be observed.

$u_i >> w_i \tan(\Delta\alpha/2)$ may have to be very strong before omission of the second term in the lower right hand corner of (38) is justified. For the large signal to noise ratios of primary interest when one is attempting to resolve closely spaced targets, this problematic region is confined to $\Delta\alpha$ very small compared to a beamwidth and therefore of limited practical concern (see later examples).

If (39) is valid - and we shall assume this for the remainder of our discussion - performance is characterized completely by the performance of the equivalent line array. Without loss of generality we can then confine our attention to line arrays broadside to the average line of sight.

5. Bounds for the Line Array

We must evaluate the diagonal elements of (39) using (27) and (28). We have postulated that the elements of the underlying two dimensional array maintain pairwise symmetry about the origin. It follows that the equivalent line array has pairs of sensors in locations $\pm u_i$. Hence

$$\sum_{i=1}^{M} u_i = 0 \tag{40}$$

$$\sum_{i=1}^{M} \underline{u}^T \Lambda_1^2 \underline{1} = \sum_{i=1}^{M} u_i \cos^2 \frac{\omega\Delta_i}{2} = 0 \tag{41}$$

$$\sum_{i=1}^{M} \underline{u}^T \Lambda_2^2 \underline{1} = \sum_{i=1}^{M} u_i \sin^2 \frac{\omega\Delta_i}{2} = 0 \tag{42}$$

Thus from (27)

$$\underline{u}^T J_t \underline{u} = 4\omega^2 \left\{ \frac{M(\frac{S}{N})^2 (1+\xi)}{1+M\frac{S}{N}(1+\xi)} \underline{u}^T \Lambda_1^2 \underline{u} + \frac{M(\frac{S}{N})^2 (1-\xi)}{1+M\frac{S}{N}(1-\xi)} \underline{u}^T \Lambda_2^2 \underline{u} \right.$$

$$\left. - \frac{4(\frac{S}{N})^2 (1+M\frac{S}{N})}{[1+M\frac{S}{N}(1+\xi)][1+M\frac{S}{N}(1-\xi)]} (\underline{u}^T \Lambda_1 \Lambda_2 \underline{1})^2 \right\} \qquad (43)$$

and from (28)

$$\underline{u}^T J_\Lambda \underline{u} = \omega^2 \left\{ \frac{(\frac{S}{N})^2}{1+M\frac{S}{N}(1+\xi)} [M(1+\xi)\underline{u}^T \Lambda_2^2 \underline{u} + 2\frac{1-M\frac{S}{N}(1+\xi)}{1+M\frac{S}{N}(1+\xi)} (\underline{u}^T \Lambda_1 \Lambda_2 \underline{1})^2] \right.$$

$$\left. + \frac{(\frac{S}{N})^2}{1+M\frac{S}{N}(1-\xi)} [M(1-\xi)\underline{u}^T \Lambda_1^2 \underline{u} + 2\frac{1-M\frac{S}{N}(1-\xi)}{1+M\frac{S}{N}(1-\xi)} (\underline{u}^T \Lambda_1 \Lambda_2 \underline{1})^2] \right\} \qquad (44)$$

The amplitude pattern ξ is given by

$$\xi = \frac{1}{M} \underline{a}_1^H \underline{a}_2 = \frac{1}{M} \sum_{i=1}^{M} \cos \omega\Delta_i \qquad (45)$$

It is now a simple matter to express the quadratic and bilinear forms in (27) in terms of ξ and its first two partial derivatives with respect to $\Delta\alpha$. Using the symbols $\xi' = \partial\xi/\partial\Delta\alpha$, $\xi'' = \partial^2\xi/\partial\Delta\alpha^2$, $\xi_o'' = \partial^2\xi/\partial\Delta\alpha^2|\Delta\alpha = 0$ one obtains

$$\frac{1}{c^2} \underline{u}^T J_t \underline{u} = \frac{4(M\frac{S}{N})^2}{r} \{M\frac{S}{N}[- \xi_o''(1-\xi^2) - \xi'^2] - \xi_o'' - \xi\xi'' - \xi'^2\} \qquad (46)$$

$$F_{\alpha_{av},\Delta\alpha} = R\, F_{\underline{t},\underline{\Delta}}\, R^T \tag{32}$$

where

$$R = \begin{bmatrix} \frac{\partial t_1}{\partial \alpha_{av}} & \cdots\cdots & \frac{\partial t_M}{\partial \alpha_{av}} & \frac{\partial \Delta_1}{\partial \alpha_{av}} & \cdots\cdots & \frac{\partial \Delta_M}{\partial \alpha_{av}} \\ \frac{\partial t_1}{\partial \Delta\alpha} & \cdots\cdots & \frac{\partial t_M}{\partial \Delta\alpha} & \frac{\partial \Delta_1}{\partial \Delta\alpha} & \cdots\cdots & \frac{\partial \Delta_M}{\partial \Delta\alpha} \end{bmatrix} \tag{33}$$

The elements of R are determined entirely by source-array geometry. In compact notation

$$R = -\frac{1}{c} \begin{bmatrix} (\cos \frac{\Delta\alpha}{2})\, \underline{u}^T & (2 \sin \frac{\Delta\alpha}{2}) \underline{w}^T \\ (\frac{1}{2} \sin \frac{\Delta\alpha}{2}) \underline{w}^T & (\cos \frac{\Delta\alpha}{2}) \underline{u}^T \end{bmatrix} \tag{34}$$

If the coordinates of the ith sensor are (x_i, y_i) the components of $\underline{u}$ and $\underline{w}$ are

$$u_i = -x_i \cos \alpha_{av} + y_i \sin \alpha_{av} \tag{35}$$

$$w_i = x_i \sin \alpha_{av} + y_i \cos \alpha_{av} \tag{36}$$

Thus $\underline{u}$ and $\underline{w}$ dependent only on average bearing. It is apparent that R becomes near-diagonal as $\Delta\alpha \to 0$. For future reference we note also that

$$\frac{\partial \underline{u}}{\partial \alpha_{av}} = \underline{w} \quad , \quad \frac{\partial \underline{w}}{\partial \alpha_{av}} = -\underline{u} \tag{37}$$

The general expression for $F_{\alpha_{av}, \Delta\alpha}$ is therefore

To gain more quantitative insight we consider a beam pattern which varies smoothly over the main lobe and can therefore be represented by a few terms of a Taylor series about the origin. Eq. (45) indicates that ξ is an even function of $\Delta\alpha$ so that the appropriate Taylor series is

$$\xi = 1 + \frac{1}{2}\,\xi_o''(\Delta\alpha)^2 + \frac{1}{24}\,\xi_o^{iv}(\Delta\alpha)^4 \;\ldots. \tag{52}$$

We shall find that terms up to order $(\Delta\alpha)^4$ are sufficient for our purposes.

A straightforward computation from (45) relates the Taylor coefficients to geometrical parameters [13]

$$\xi = 1 - 2\pi^2\,E_S(\Delta\alpha)^2 + \frac{2\pi^4}{3}\,D_S(\Delta\alpha)^4 \tag{53}$$

If λ is the wavelength of the narrowband signal

$$E_S = \frac{1}{M}\sum_{i=1}^{M}\left(\frac{u_i}{\lambda}\right)^2 \tag{54}$$

$$D_S = \frac{1}{M}\sum_{i=1}^{M}\left(\frac{u_i}{\lambda}\right)^4 \tag{55}$$

To the extent of our approximation E_S and D_S completely describe the relevant features of array geometry.

The leading (SNR dependent) term of (46) and (47) can now be written in the form

$$M\,\frac{S}{N}\,[-\,\xi_o''(1-\xi^2) - \xi'^2] = 16\pi^6\,M\,\frac{S}{N}\,E_S(D_S-E_S^{\,2})(\Delta\alpha)^4 \tag{56}$$

Similarly

$$-\,\xi_o'' - \xi\xi'' - \xi'^2 = 8\pi^2[E_S-\pi^2(D_S+3E_S^{\,2})(\Delta\alpha)^2] \tag{57}$$

To deal with the r dependent terms of (47) and (48) it is convenient to distinguish between two ranges of signal to noise ratio

a) When $M\frac{S}{N}(1-\xi^2) \gg 2$ (hence surely $M\frac{S}{N} \gg 1$)

$$\frac{(M\frac{S}{N})^2}{r} \approx \frac{1}{1-\xi^2} \tag{58}$$

b) When $M\frac{S}{N}(1-\xi^2) \ll 2$

$$\frac{(M\frac{S}{N})^2}{r} \approx \frac{(M\frac{S}{N})^2}{1+2M\frac{S}{N}} \tag{59}$$

In case A) (which can extend well into the main lobe of the beampattern for large SNR)

$$-\xi_o'' + \xi\xi'' + \xi'^2 + \frac{4(M\frac{S}{N})^2}{r}(\xi\xi')^2$$

$$\approx 8\,{}^4(D_S+3E_S^2)(\Delta\alpha)^2 + 16\pi^2\,\frac{[E_S + \frac{2\pi^2}{3}(D_S+3E_S^2)(\Delta\alpha)^2]^2}{E_S - \frac{\pi^2}{3}(D_S+3E_S^2)(\Delta\alpha)^2} \tag{60}$$

and

$$\frac{(M\frac{S}{N})^2}{r} \approx \frac{1}{1-\xi^2} = \frac{1}{4\pi^2(\Delta\alpha)^2[E_S - \frac{\pi^2}{3}(D_S+3E_S^2)(\Delta\alpha)^2]} \tag{61}$$

In the above approximations we have retained terms up to order $(\Delta\alpha)^2$ except for (56) where the lowest order term [of order $(\Delta\alpha)^4$] cannot be ignored because it is multiplied by $M\frac{S}{N}$.

We know that $1-\xi^2 \geq 0$. It is therefore apparent that $E_S \gg \frac{\pi^2}{3}(D_S+3E_S^2)(\Delta\alpha)^2$ throughout the region where our approximation is valid. As $\Delta\alpha$ decreases to values well within the main beam, this inequality becomes stronger and we can ignore the terms in $(D_S+3E_S^2)(\Delta\alpha)^2$. With this approximation Eqs. (46) and (47) yield

$$\frac{1}{c^2}\underline{u}^T J_t \underline{u} \approx \frac{16}{(\Delta\alpha)^2}\left[\frac{1}{2}+\pi^4 M\frac{S}{N}(D_S-E_S^2)(\Delta\alpha)^4\right] \tag{62}$$

$$\frac{1}{c^2}\underline{u}^T J_\Delta \underline{u} \approx \frac{4}{(\Delta\alpha)^2}\left[1+\pi^4 M\frac{S}{N}(D_S-E_S^2)(\Delta\alpha)^4\right] \tag{63}$$

When the second terms of (62) and (63) dominate, both quantities decrease with decreasing $(\Delta\alpha)$. For small $\Delta\alpha$ the first terms dominate and both quantities increase with decreasing $\Delta\alpha$. The Cramer-Rao bounds on α_{av} and $\Delta\alpha$ [proportional to $(\underline{u}^T J_t \underline{u})^{-1}$ and $(\underline{u}^T J_\Delta \underline{u})^{-1}$ respectively] therefore exhibit a peak at a value of $\Delta\alpha$ easily calculated. For α_{av} this peak occurs at

$$\Delta\alpha_p = \left[2M\frac{S}{N}\pi^4(D_S-E_S^2)\right]^{-\frac{1}{4}} \tag{64}$$

The corresponding peak for $\Delta\alpha$ occurs at $\Delta\alpha_q = 2^{1/4}\Delta\alpha_p$. For $\Delta\alpha \gg \Delta\alpha_q$ the ratio of the two Cramer-Rao bounds is fixed at 4, the same as the ratio for widely separated sources. For $\Delta\alpha \ll \Delta\alpha_q$ the ratio gradually decreases to 2.

It is interesting to observe that array geometry enters into (62) and (63) only through the factor $D_S-E_S^2$, which can be interpreted as the sample variance of the u_i^2/λ^2 [see (54) and (55)]. It is a non-negative quantity which goes to zero when half of the sensors are placed at each end of the available aperture. This is precisely the geometry which maximizes E_S (hence $-\xi_o''$) and therefore minimizes the mean square bearing error for a single source or for widely separated sources. Thus the requirements on array geometry change when one is attempting to separate sources within the main lobe of the beam pattern.

In regime B $[M \frac{S}{N}(1-\xi^2) << 2]$ r is a constant and (46) converges to

$$\frac{1}{c^2} \underline{u}^T J_t \underline{u} = \frac{32\pi^2 (M \frac{S}{N})^2}{1+2M \frac{S}{N}} E_S \tag{65}$$

Since $\xi_o'' = -4\pi^2 E_S$ this is simply the double strength single target result of (50). The dominant terms of (47) now vary as $(\Delta\alpha)^2$ so that the Cramer-Rao bound for $\Delta\alpha$ diverges with $(\Delta\alpha)^{-2}$. It follows that the Cramer-Rao bound for $\Delta\alpha$ has a minimum near the point where $M \frac{S}{N} (1-\xi^2) = 2$.

The bounds are particularly simple when the post-beamforming signal to noise ratio is low. In that case (46) and (47) become

$$\frac{1}{c^2} \underline{u}^T J_t \underline{u} \quad 4(M \frac{S}{N})^2 [-\xi_o'' - (\xi\xi'' + \xi'^2)] \approx 32\pi^2 (M \frac{S}{N})^2 [E_S - \pi^2 (D_S + 3E_S{}^2)(\Delta\alpha)^2] \tag{66}$$

$$\frac{1}{c^2} \underline{u}^T J_\Delta \underline{u} \quad (M \frac{S}{N})^2 [-\xi_o'' + (\xi\xi'' + \xi'^2)] \approx 8\pi^4 (M \frac{S}{N})^2 (D_S + 3E_S{}^2)(\Delta\alpha)^2 \tag{67}$$

ξ'^2 reaches a maximum at the inflection point $\Delta\alpha_i$ of the pattern function, where $\xi'' = 0$. Hence $\xi\xi'' + \xi'^2$ has a maximum near $\Delta\alpha_i$ with the result that the bound for α_{av} exhibits a local maximum, the bound for $\Delta\alpha$ a local minimum. Within the range of applicability of the Taylor expansion, the bound for α_{av} decreases steadily with decreasing $\Delta\alpha$, while the bound for $\Delta\alpha$ increases steadily.

Fig. 4 shows numerical results for a 30 element line array with regular half wavelength spacing, $\alpha_{av} = 0$. The half-width of the main lobe of ξ is approximately 4°. The upper set of curves illustrates the case of high MS/N, the lower set the case of small M S/N. All computation were made from (47) and (48), i.e. without approximation. The range of validity of the Taylor expansion is approximately $\Delta\alpha < 3^o$. Within that region all features of the

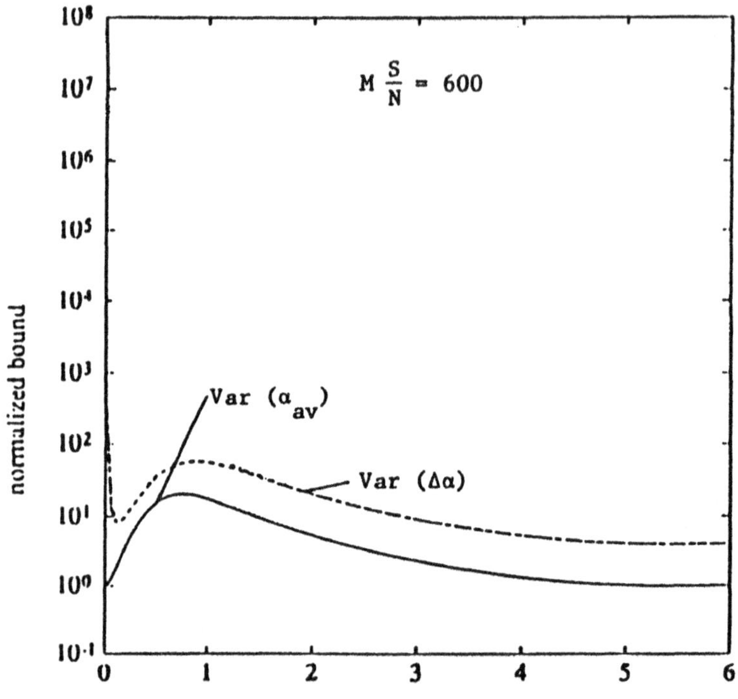

M S/N = 600
Var (α_av)
Var (Δα)
normalized bound
angular separation in degrees

Fig. 4a

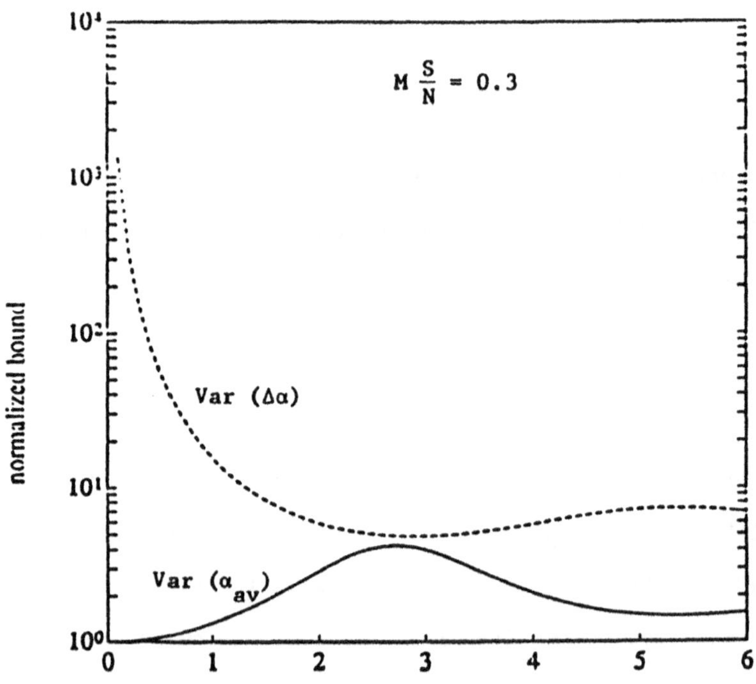

M S/N = 0.3
Var (Δα)
Var (α_av)
normalized bound
angular separation in degrees

Fig. 4b

curves agree well with the analytical predictions.

6. Effect of Unknown Spectral Levels

In most practical cases it would be unrealistic to assume that the power levels of the two signals are known a priori. If they are not, they must be included as additional (nuisance) parameters in the estimation problem. The vector of unknown parameters is now $\phi = (\alpha_{av}, \Delta\alpha, S_1, S_2)^T$. To avoid excessive algebraic complexity we shall concentrate on the case of signal powers which happen to be equal ($S_1 = S_2 = S$), although this is not known a priori.

It is convenient to partition the Fisher Information for ϕ into four 2x2 matrices.

$$F_{\underline{\phi}} = \left[\begin{array}{c|c} F_{\alpha_{av},\Delta\alpha} & F_{\underline{\alpha S}} \\ \hline F_{\underline{S\alpha}} & F_{\underline{S}} \end{array}\right] \tag{68}$$

Under the previously stated restrictions on array geometry (which allow use of the average projected array), $F_{\alpha_{av},\Delta\alpha}$ is given by (39), (46) and (47). Under the same constraints the remaining entries of (68) are (see [13] for computational details)

$$F_{\alpha S} = F^T_{S\alpha} = \frac{TW}{2\pi}\left(\frac{M}{N}\right)^2 \frac{S}{r}\,\xi\zeta' \begin{bmatrix} -2 & 2 \\ \frac{1}{r}[1-(M\frac{S}{N})^2(1-\xi^2)] & \frac{1}{r}[1-(M\frac{S}{N})^2(1-\xi^2)] \end{bmatrix} \tag{69}$$

and

$$F_S = \frac{TW}{2\pi}\left(\frac{M}{Nr}\right)^2 \begin{bmatrix} [1+M\frac{S}{N}(1-\zeta^2)]^2 & \zeta^2 \\ \zeta^2 & [1+M\frac{S}{N}(1-\zeta^2)]^2 \end{bmatrix} \tag{70}$$

It is immediately apparent from (69) that $\underline{F}_{\alpha S} = 0$ whenever $\zeta = 0$, which is true for source separations large compared with the beamwidth. In that case one is dealing effectively with the single source localization problem for which location and spectral estimation errors are known to be uncoupled. When both sources lie within the same beam, the coupling can be significant and its effect must be studied. The fact that $\underline{F}_{\alpha S}$ also vanishes at $\Delta\alpha = 0$ is deceptive because F_S becomes singular at that point.

The Cramer-Rao bound on $(\Lambda_{av}, \Delta\alpha)$ is given by the first two rows and columns of $F_{\underline{\phi}}^{-1}$. From (68) these are

$$\mathrm{CRLB}(\alpha_{av}, \Delta\alpha) = [F_{\alpha_{av},\Delta\alpha} - \underline{F}_{\alpha S}\,\underline{F}_S^{-1}\,\underline{F}_{\alpha S}^T]^{-1} \tag{71}$$

A straightforward computation yields

$$\Delta F \doteq \underline{F}_{\alpha S}\,\underline{F}_S^{-1}\,\underline{F}_{\alpha S}^T = \frac{TW}{2\pi}\left(2M\frac{S}{N}\zeta\zeta'\right)^2 \begin{bmatrix} \dfrac{2}{r(1-\zeta^2)} & 0 \\ 0 & \dfrac{[1-(M\frac{S}{N})^2(1-\zeta^2)]^2}{2r^2\{[1+M\frac{S}{N}(1-\zeta^2)]^2+\zeta^2\}} \end{bmatrix} \tag{72}$$

Since $F_{\alpha_{av},\Delta\alpha}$ is also diagonal, it follows that the estimation errors of average and differential bearing remain uncoupled, even when the signal power levels are unknown. We note that this simple diagonal form does not occur, in general, when the signal power levels happen to be unequal.

The fact that $\Delta F = 0$ when $\xi = 0$ confirms the decoupling of location and spectral estimates for widely separated sources. At the opposite extreme, when $\Delta\alpha \to 0$ and hence $\xi \to 1$, $\xi' \to 0$ one finds after a few lines of algebra

$$\Delta F = \frac{TW}{2\pi} \begin{bmatrix} \dfrac{8(M\frac{S}{N})^2}{1+2M\frac{S}{N}} (-\xi_o'') & 0 \\ 0 & 0 \end{bmatrix} \tag{73}$$

The non-zero entry is equal to the Fisher Information of α_{av} for known power levels [see (50)]. Hence the Cramer-Rao bound on α_{av} now tends to infinity as $\Delta\alpha \to 0$ (in contrast with the case of known power levels). The Cramer-Rao bound on $\Delta\alpha$ tended to zero even for known power levels and will clearly continue to do so when the power levels are unknown.

There are two remaining questions of interest: 1) where does the α_{av} bound for unknown power levels begin to deviate significantly from the known power curve? and 2) Is there any range of $\Delta\alpha$ where the bounds on source separation differ significantly in the known and unknown power cases?

To gain insight into 1) we combine (46) with the first line of (72) to obtain the Fisher Information of α_{av} for unknown power levels.

$$F_{\alpha_{av}} = \frac{TW}{2\pi} \frac{4(M\frac{S}{N})^2}{r} \{ M\frac{S}{N} [-\xi_o''(1-\xi^2)-\xi'^2]-\xi_o'' - \xi\xi''-\xi'^2 - \frac{2(\xi\xi')^2}{1-\xi^2}\} \tag{74}$$

Using Taylor series expansions as in Section 5 it is a simple matter to demonstrate that

$$-\xi_o''-\xi\xi''-\xi'^2-\frac{2(\xi\xi')^2}{1-\xi^2} = 0 \quad \text{to order } (\Delta\alpha)^2 \tag{75}$$

The entire term in { } in (74) therefore varies as $(\Delta\alpha)^4$ [see (56)], so that the bound on α_{av} rises steadily with decreasing $\Delta\alpha$. When $M\frac{S}{N} \gg 1$ the term in $(M\frac{S}{N})$ dominates (74). In the known power case the corresponding term dominated for decreasing $\Delta\alpha$ until $\Delta\alpha$ reached a point near the peak of the α_{av} curve [see Fig. 4a]. This point therefore marks the point of departure for the unknown power curve, which simply continues to rise as $\Delta\alpha$ decreases. When $M\frac{S}{N} \ll 1$ the term in $M\frac{S}{N}$ can be ignored and (74) indicates that the known and unknown power curves separate when $(\xi\xi')^2/(1-\xi^2)$ begins to vary significantly, which is at the edge of the beam pattern.

To address the second question we rewrite the $\Delta\alpha$ component of ΔF in (72) in the form

$$\Delta F_{\Delta\alpha} = \frac{TW}{2\pi}\,\frac{4(M\frac{S}{N})^4}{r^2}\,(\xi\xi')^2\,\frac{1}{2}\,\frac{\left[(M\frac{S}{N})(1-\xi^2)-\frac{1}{M\,S/N}\right]^2}{[M\frac{S}{N}(1-\xi^2)+1]^2+\xi^2} \tag{76}$$

The corresponding term in $F_{\Delta\alpha}$ is from (47)

$$\frac{TW}{2\pi}\,\frac{4(M\frac{S}{N})^4}{r^2}\,(\xi\xi')^2 \tag{77}$$

For all $M\frac{S}{N} \geq 1$

$$\frac{[(M\frac{S}{N})(1-\xi^2) - \frac{1}{M\,S/N}]^2}{[(M\frac{S}{N})(1-\xi^2) + 1]^2 + \xi^2} \leq 1 \tag{78}$$

It follows that the bounds on $\Delta\alpha$ for known and unknown power levels can differ by no more than 3db as long as $MS/N \geq 1$. For most combinations of parameter values the difference will clearly be smaller.

The argument - but not the conclusion - is somewhat different for low post-beamforming signal to noise ratios. For known power, from (67)

$$F_{\Delta\alpha} = \frac{TW}{2\pi}(M\frac{S}{N})^2[-\xi_o''+\xi\xi''+\xi'] \approx \frac{TW}{2\pi}8\pi^4(M\frac{S}{N})^2(D_S+3E_S^2)(\Delta\alpha)^2 \tag{79}$$

But from (76)

$$\Delta F_{\Delta\alpha} = \frac{TW}{2\pi}2(M\frac{S}{N})^2\frac{(\xi\xi')^2}{1+\xi^2} \approx \frac{TW}{2\pi}16\pi^4(M\frac{S}{N})^2 E_S^2(\Delta\alpha)^2 \tag{80}$$

It follows that for unknown power the Fisher Information of $\Delta\alpha$ is

$$F_{\Delta\alpha} = \frac{TW}{2\pi}(M\frac{S}{N})^2[-\xi_o''+\xi\xi''+\xi'^2(1-\frac{2\xi^2}{1+\xi^2})] \approx \frac{TW}{2\pi}8\pi^4(M\frac{S}{N})^2(D_S+E_S^2)(\Delta\alpha)^2 \tag{81}$$

Since $D_S \geqslant E_S^2$ the expressions in (79) and (81) can differ by at most 3db.

Fig. 5 shows curves for both known and unknown signal power for the same regular line array as in Fig. 4. As expected, the primary difference between the known and unknown power cases concerns the bounds on α_{av}. The unknown power curves rise steadily with decreasing $\Delta\alpha$, deviating from the known power curves below the values of $\Delta\alpha$ identified in the analysis. There is little difference between the two bounds on $\Delta\alpha$.

7. Uncertainties in Array Geometry

When large arrays are deployed in the ocean, locations of individual sensors are generally not known with great precision. This section is concerned with the effect of such uncertainties on the estimates of α_{av} and $\Delta\alpha$.

The nominal geometry is the same as shown in Fig. 1. In reality each sensor is located at a point displaced from its nominal location by a distance typically small compared with the intersensor spacing. We designate displacements of the ith sensor along the average line of sight and at right angles to it by Δw_i and Δu_i respectively. We assume that $(\Delta u_1 \ldots \Delta u_M, \Delta w_1 \ldots \Delta w_M)$ are independent random variables with probability densities.

$$p(\Delta u_i) = \frac{1}{\sqrt{2\pi}\,\sigma} \exp\left[-\frac{(\Delta u_i)^2}{2\sigma^2}\right] \quad i = 1 \ldots M \tag{82}$$

$$p(\Delta w_i) = \frac{1}{\sqrt{2\pi}\,\sigma} \exp\left[-\frac{(\Delta w_i)^2}{2\sigma^2}\right] \quad i = 1 \ldots M \tag{83}$$

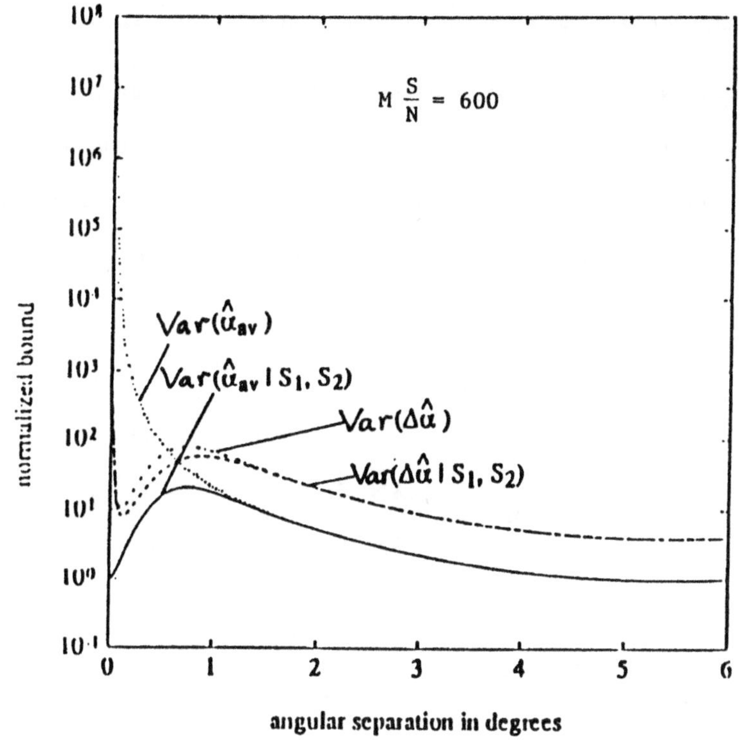
M S/N = 600
Var(α̂av)
Var(α̂av | S1, S2)
Var(Δα̂)
Var(Δα̂ | S1, S2)
normalized bound
angular separation in degrees

Fig. 5a

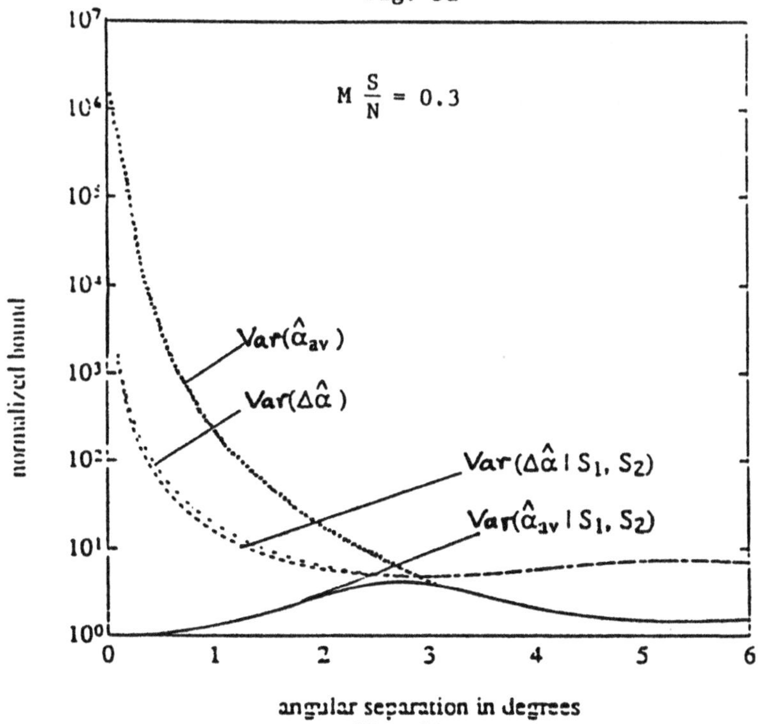
M S/N = 0.3
Var(α̂av)
Var(Δα̂)
Var(Δα̂ | S1, S2)
Var(α̂av | S1, S2)
normalized bound
angular separation in degrees

Fig. 5b

The standard deviation σ is small compared with the intersensor spacing.

The Δu_i and Δw_i are now nuisance parameters which must be estimated along with α_{av} and $\Delta\alpha$. Since prior statistics are available for the nuisance parameters, but not for α_{av} and $\Delta\alpha$, we use the hybrid version of the Cramer-Rao bound [14]. If the data vector is $\underline{x}$ and the vector of parameters is $\underline{\phi} = (\alpha_{av}, \Delta\alpha, \Delta\underline{u}^T, \Delta\underline{w}^T)^T$, the Fisher Information is

$$F_{\underline{\phi}} = F_1 + F_2 \tag{84}$$

$$F_1 = -E_{\underline{x},\Delta u,\Delta \underline{w}} \left\{ \frac{\partial^2 p(\underline{x}/\phi)}{\partial\phi_i \partial\phi_j} \right\} \tag{85}$$

$$F_2 = -E_{\Delta\underline{u},\Delta\underline{w}} \left\{ \frac{\partial^2 p(\Delta\underline{u},\Delta\underline{w})}{\partial\phi_i \partial\phi_j} \right\} \tag{86}$$

Thus F_1 is the usual Fisher Information of $\underline{\phi}$ averaged over the known prior distribution, while F_2 characterizes the information furnished by the prior distribution. The desired Cramer-Rao bound is F_{ϕ}^{-1}, whose diagonal elements are the mean square errors of the various components of $\underline{\phi}$, here averaged over the prior distributions of $\Delta\underline{u}$ and $\Delta\underline{w}$.

It is clear that displacement of sensor i by a distance small compared with the intersensor spacing will have negligible effect on the differential signal travel time $\Delta_i = \tau_i - \sigma_i$. Since the Fisher Information for $\underline{t}$ and $\underline{\Delta}$ depends only on the Δ_i, Eqs. (27) and (28) remain valid. The uncertainties become important only when one attempts to convert measured delays and differential delays into location parameters for sources and sensors. The transformation matrix equivalent to R in (32) is now an (2M+2)x2M matrix, since the vector of unknown parameters has dimension

2M+2. It is immediately obvious from dimensional considerations that $F_{\underline{\phi}}$ is singular in the absence of prior information concerning $\Delta\underline{u}$ and $\Delta\underline{w}$.

Considerations entirely parallel to those developed in Section 4 allow one to project the nominal array onto a straight line at right angles to the average line of sight. The same restrictions on array geometry as in Section 4 are, of course, implied. It is intuitively obvious (and not difficult to prove formally) that small displacements of the projected sensors at right angles to the line of sight have negligible effect on the signal travel times. In summary, therefore, one can always work with a line array broadside to the average line of sight and consider sensor displacements only in the direction of the line of sight (Δw_i). The effective parameter vector is now $(\alpha_{av},\Delta\alpha,\Delta\underline{w}^T)^T$ and the dimension of R is (M+2)xM.

The procedure for calculating the Cramer-Rao bound on $(\alpha_{av},\Delta\alpha)$ is quite analoguous to that used in dealing with the nuisance parameter $\underline{S}$ in Section 6. The Fisher Information $F_{\underline{\phi}}$ for the complete parameter vector $\underline{\phi}$ is now given by (84). F_2 is the (M+2)x(M+2) matrix

$$F_2 = \frac{TW}{2\pi}\left[\begin{array}{cc|ccc} 0 & 0 & 0 & \cdots & 0 \\ 0 & 0 & 0 & \cdots & 0 \\ \hline 0 & 0 & & & \\ \vdots & \vdots & & \frac{1}{\sigma^2}I & \\ 0 & 0 & & & \end{array}\right] \qquad (87)$$

The two rows and columns of zeros result from the fact that the prior distribution of $\Delta\underline{w}$ is not a function of α_{av} or $\Delta\alpha$.

Partitioning $F_{\underline{\phi}}$ as in (68) one can write

$$F_{\underline{\phi}} = \left[\begin{array}{c|c} F_{\alpha_{av},\Delta\alpha} & F_{\alpha\underline{\Delta w}} \\ \hline F_{\underline{\Delta w}\alpha} & F_{\underline{\Delta w}} \end{array} \right] \tag{88}$$

The equivalent of (71) is now

$$\mathrm{CRLB}(\alpha_{av},\Delta\alpha) = [F_{\alpha_{av},\Delta\alpha} - F_{\alpha\underline{\Delta w}} F_{\underline{\Delta w}}^{-1} F_{\alpha\underline{\Delta w}}^{T}]^{-1} \tag{89}$$

For the small sensor displacements assumed throughout this discussion the averaging operation of F_1 over $\Delta\underline{w}$ is trivial, so that $F_{\Lambda_{av},\Delta\alpha}$ retains the value given by (39). A tedious but straightforward computation (see [13]) gives the change due to sensor displacements.

$$\Delta F = F_{\alpha\underline{\Delta w}} F_{\underline{\Delta w}}^{-1} F_{\alpha\underline{\Delta w}}^{T}$$

$$= \frac{TW}{2\pi c^2} \cos^2\frac{\Delta\alpha}{2} \begin{bmatrix} \underline{u}^T J_t (J_t + \frac{c^2}{\sigma^2 \cos^2\frac{\Delta\alpha}{2}} I)^{-1} J_t \underline{u} & 0 \\ 0 & 0 \end{bmatrix} \tag{90}$$

Thus sensor location uncertainty degrades the accuracy of the α_{av} estimate but has no effect on the $\Delta\alpha$ estimate.[2] To gain quantitative insight into the incremental α_{av} error we consider the fractional change in the Cramer-Rao bound. For $\Delta\alpha \to 0$ the computations are easily performed because J_t is now given by the simple expression of (30). After a few lines of algebra one obtains

---

[2] More careful analysis, carrying $\Delta\underline{u}$ as well as $\Delta\underline{w}$ displacements, shows that the $\Delta\underline{u}$ displacements affect the $\Delta\alpha$ estimate but not the α_{av} estimate. As anticipated however, the effect on the $\Delta\alpha$ estimate is small as long as $\Delta\alpha$ is small. (See [13] for details).

$$\frac{\Delta CRLB(\alpha_{av})}{CRLB\alpha_{av}\big|_{\Delta \underline{w}=0}} = 8\pi^2 \left(\frac{\sigma}{\lambda}\right)^2 \frac{1}{M} \frac{\left(2M \frac{S}{N}\right)^2}{1+2M \frac{S}{N}} \tag{91}$$

It is obvious that (91) can easily exceed unity, even for sensor location uncertainties smaller than the wavelength λ. The SNR dependence of (91) results from the fact that CRLB $\alpha_{av}\big|_{\Delta \underline{w} = 0}$ decreases with SNR in the indicated fashion. The increment ΔCRLB(α_{av}) is independent of SNR. For $\Delta\alpha > 0$ (but small enough to keep both sources within the same beam) it is not difficult to establish [13] that the right side of (91) gives an upper bound on the fractional change of CRLB(α_{av}).

There is a simple physical explanation for the dependence of the bounds on sensor location uncertainty. Sensor displacements in the w direction tilt the baseline of the two element array formed by any pair of sensors. Even if the differential delay measurement were perfect, the calculated bearing for each source would therefore be in error. However, for small $\Delta\alpha$ the errors in the two bearing measurements would be nearly the same, with the result that there is no incremental error in the $\Delta\alpha$ measurement but an increase close to the single source increment in the α_{av} measurement. We should point out that this particular conclusion depends heavily on the assumption of equal source power. In the more general problem there exist linear combinations of the two bearings which have the indicated properties, but they would not be the simple average and difference.

8. Unequal Signal Power Levels

We conclude our discussion with some remarks concerning the Cramer-Rao bounds on $\Delta\alpha$ when the signal power levels are unequal. In the interest of algebraic simplicity we concentrate on a situation at the opposite extreme from that considered thus far: Signals whose power levels are very unequal. The question of primary interest is whether one can make meaningful statements concerning the distance separating a weak source from a nearby strong source.

We take source 1 as the strong source and assume $MS_1/N >> 1$. For the weak source we assume $MS_2/N << 1$. As before, we work with the line array generated by projecting the actual array onto a straight line at right angles to the average line of sight. After considerable algebraic manipulation [13] one obtains the following expression for the Fisher Information of the source bearings α_1 and α_2

$$F_{\underline{\alpha}} = \begin{bmatrix} F_{\alpha_1} & F_{\alpha_1\alpha_2} \\ F_{\alpha_1\alpha_2} & F_{\alpha_2} \end{bmatrix} \tag{92}$$

where

$$F_{\alpha_1} = \frac{TW}{2\pi} \frac{2(M\frac{S_1}{N})^2}{1+M\frac{S_1}{N}} (-\xi_o'') \tag{93}$$

$$F_{\alpha_1\alpha_2} = \frac{TW}{2\pi} \frac{2(M\frac{S_1}{N})(M\frac{S_2}{N})}{1+M\frac{S_1}{N}} (-\xi\xi''-\xi'^2) \tag{94}$$

$$J_{\alpha_2} = \frac{TW}{2\pi} \frac{2(M\frac{S_2}{N})^2}{1+M\frac{S_1}{N}} \{-\xi_o''+M\frac{S_1}{N}[-\xi_o''(1-\xi^2)+\xi'^2]\} \tag{95}$$

As expected, the α_1 and α_2 estimates are strongly coupled. Our main concern is with $\alpha_1-\alpha_2$. We therefore seek a linear transformation

$$\Delta\alpha = \alpha_1-\alpha_2 \tag{96}$$

$$z = k_1\alpha_1+k_2\alpha_2$$

which will yield uncorrelated estimates for $\Delta\alpha$ and z. The Fisher Information for $(\Delta\alpha, z)$ is

$$F_{\Delta\alpha,z} = \frac{1}{(k_1+k_2)^2} \begin{bmatrix} 1 & 1 \\ k_2 & -k_1 \end{bmatrix} \cdot \begin{bmatrix} F_{\alpha_1} & F_{\alpha_1\alpha_2} \\ F_{\alpha_1\alpha_2} & F_{\alpha_2} \end{bmatrix} \cdot \begin{bmatrix} 1 & k_2 \\ 1 & -k_1 \end{bmatrix} \tag{97}$$

This matrix is diagonal when

$$k_2 = k_1 \frac{F_{\alpha_2} + F_{\alpha_1\alpha_2}}{F_{\alpha_1} + F_{\alpha_1\alpha_2}} \tag{98}$$

For $M\frac{S_1}{N} >> 1 >> M\frac{S_2}{N}$ it is obvious from (93)-(95) that $F_{\alpha_1} >> F_{\alpha_1\alpha_2} >> F_{\alpha_2}$ so that (98) becomes

$$k_2 = k_1 \frac{F_{\alpha_1\alpha_2}}{F_{\alpha_1}}$$

With that choice

$$F_{\Delta\alpha} = F_{\alpha_2} - \frac{F_{\alpha_1\alpha_2}}{F_{\alpha_1}}$$

$$\approx \frac{TW}{\pi}\left(M\frac{S_2}{N}\right)^2 \{-\xi_o''(1-\xi^2)+\xi'^2(2\xi^2-1)\} \tag{99}$$

$F_{\Delta\alpha}$ depends only on the SNR of the weak signal. $-\xi_o''(1-\xi^2)$ decreases steadily with $\Delta\alpha$ while $\xi'^2(2\xi^2-1)$ exhibits a maximum near the inflection point $\Delta\alpha_i$ of the beampattern. For $\Delta\alpha << \Delta\alpha_i$ the Taylor expansion to order $(\Delta\alpha)^2$ yields

$$F_{\Delta\alpha} \approx \frac{TW}{\pi}\left(M\frac{S_2}{N}\right)^2 \cdot 32\pi^4 E_S^2(\Delta\alpha)^2 \tag{100}$$

Thus the Cramer-Rao bound varies as $(\Delta\alpha)^{-2}$, as in the equal power case treated in Section 4. Fig. 6 shows the Cramer-Rao bound on $\Delta\alpha$ for the regular 30 element array used in earlier illustrations. The $\Delta\alpha$ dependence is clearly consistent with the analytical predictions. One also finds that changes of the two signal to noise ratios have virtually no effect on the shape of the curve as long as one maintains the restrictions $MS_1/N \gg 1$, $MS_2/N \ll 1$. Perhaps the most important conclusion is that the bound is not dramatically different from the bound obtained earlier for the case of weak signals of equal strength.

Acknowledgement

The work reported in this paper was supported in part by the Office of Naval Research under Contract N00014-80-C-0092.

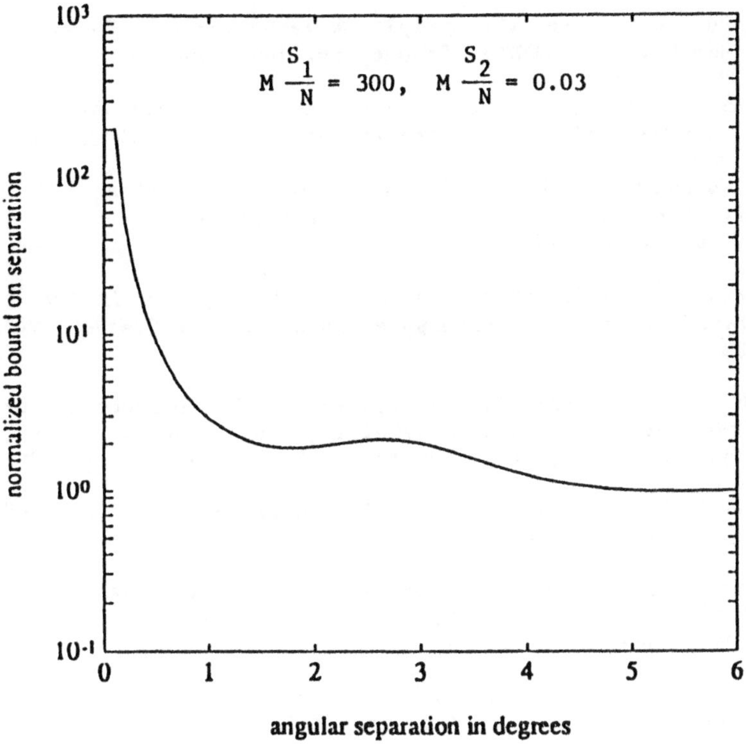

$M\frac{S_1}{N} = 300, \quad M\frac{S_2}{N} = 0.03$
normalized bound on separation
10³
10²
10¹
10⁰
10⁻¹
0
1
2
3
4
5
6
angular separation in degrees

Fig. 6

References

1. J. Capon, "High Resolution Frequency Wave Number Spectrum Analysis", Proceedings of the IEEE, Vol. 57, pp. 1408-1418, 1969.

2. G. Bienvenu and L. Kopp, "Principle de la Goniometrie Passive Adaptive", Proc. T'eme Colloque GRETSI, France, pp. 106/1-106/10.

3. R.O. Schmidt, "A Signal Subspace Approach to Multiple Emitter Location and Spectral Estimation", Ph.D. Dissertation, Stanford Univ., 1981.

4. Guaning Su and Martin Morf, "The Signal Subspace Approach for Multiple Wide-Band Emitter Location", IEEE Trans. Acoustics, Speech, Signal Processing, Vol. 31, No. 6, Dec. 1983

5. R. Kumeresan and D.W. Tufts, "Estimating the Angle of Arrival of Multiple Plane Waves", IEEE Trans. Aerospace and Electronic Systems, Vol. 19, pp. 134-139, 1983.

6. M. Kaveh and A.J. Barabell, "The Statistical Performance of the MUSIC and the Minimum-Norm Algorithms in Resolving Plane Waves in Noise", IEEE Trans. Acoustics, Speech, Signal Processing, Vol. ASSP-34, April 1986.

7. B. Hellwell, "Direction Estimation for Multiple Remote Sources", Ph.D. Dissertation, Yale Univ., 1983.

8. J.F. Bohme, "Estimation of Source Parameters by Maximum Likelihood and Nonlinear Regression", Proc. IEEE Int. Conf. on Acoustic, Speech and Signal Processing (ICASSP), 1984.

9. H. Messer and P.M. Schultheiss, "Parameter Estimation for Two Sources with Overlapping Spectra Using an Arbitrary M Sensor Array", Proceedings of the 3rd IEEE Workshop on Spectrum Estimation and Modeling, Boston, 1986.

10. P. Stoica and A. Nehorai, "MUSIC, Maximum Likelihood and Cramer-Rao Bound", Report No. 8708 of Center for Systems Science, Yale University, July 1987.

11. H.L. Van Trees, "Detection, Estimation and Modulation Theory, Part I", John Wiley and Sons, New York 1968.

12. W.J. Bangs, "Array Processing with Generalized Beamformers", Ph.D. Dissertation, Yale University, 1973.

13. A. Bhuyan, "Performance Analysis of High Resolution Passive Systems", Ph.D. Dissertation, Yale University, 1989.

14. Y. Rockah, P.M. Schultheiss, "Array Shape Calibration Using Sources in Unknown Location, Part I: Far Field Sources", IEEE Trans. ASSP-35, 286, 1987.

CHAPTER 6

RAPIDLY ADAPTIVE NULLING OF INTERFERENCE

D.W. TUFTS
University of Rhode Island (USA)

I.P. KIRSTEINS
University of Rhode Island (USA) *

Rapidly Adaptive Nulling of Interference

by

Ivars P. Kirsteins† and Donald W. Tufts

University of Rhode Island

Department of Electrical Engineering

Kingston, Rhode Island U.S.A. 02881

† on training assignment from the Naval Underwater Systems Center, Code 3314, New London, CT U.S.A. 06320

Abstract

We present a simpler and more general analysis of our previously proposed Principal Component Inverse (PCI) method of rapidly adaptive nulling of interference. We assume that the interference consists of a strong gaussian component with a rank deficient covariance matrix plus a weak component of white gaussian background noise. We derive an approximate Beta probability density function of the output signal-to-noise ratio (SNR) for the PCI method. Using our theoretically derived formulas, we show that the PCI method requires much less data to produce a given, needed level of SNR with high probability than the Sample Matrix Inverse (SMI) method based on the inverse of the sample covariance matrix. The approximations and the final probability density function are tested through computer simulation. They accurately explain the experimental results.

This research was sponsored by the SDIO/IST, managed by the Army Research Office under contract DAAL03-86-K-0108, Donald W. Tufts, principal investigator.

1.0 Introduction

Consider the problem of detecting a known signal with random phase in the presence of complex gaussian distributed noise. The two hypothesis are:

$$H0 : \mathbf{x} = \mathbf{d} \tag{1}$$

$$H1 : \mathbf{x} = \mathbf{d} + e^{j\Theta}\mathbf{s} \tag{2}$$

where $\mathbf{x}$ is the vector of complex-valued observed data, $\mathbf{d}$ is a complex gaussian distributed random vector with zero-mean and covariance matrix $E[\mathbf{d}\mathbf{d}^H] = R$, and $\mathbf{s}$ is the known shape vector of the signal to be detected. The dimensions of $\mathbf{x}$, $\mathbf{d}$, and $\mathbf{s}$ are $n \times 1$. The signal phase Θ is assumed to be uniformly distributed on $(0, 2\pi)$. This scenario typically arises in radar or sonar where the elements of the received data vector $\mathbf{x}$ would represent spatial/temporal samples of the electromagnetic or acoustic field observed at an array of sensors.

It can be shown that a likelihood ratio test statistic [1] is

$$\alpha = \left|\mathbf{s}^H R^{-1} \mathbf{x}\right| \tag{3}$$

where the operator " ^{H} " means a Hermitian, ie., complex conjugate transpose. A filtering interpretation of the likelihood ratio test statistic (3) can be gotten by rewriting formula (3) as

$$\alpha = \left|\mathbf{w}^H \mathbf{x}\right| \tag{4}$$

where the filter weights $\mathbf{w}$ are given by

$$\mathbf{w} = R^{-1}\mathbf{s} \quad . \tag{5}$$

However in practice, sufficient information for implementation of the above detector is often unavailable. The underlying physical mechanisms generating the interference may be unknown or changing rapidly. This situation commonly occurs when dealing with clutter from radar returns or reverberation in active sonar. A common approach [6] thus is to adaptively estimate the the weight vector from observed noise data.

Our objective here is to present the rationale and performance of our Principal Component Inverse (PCI) method of very fast adaptation. We begin in the next subsection by describing the earlier, important Sample Matrix Inverse (SMI) method. A drawback of the SMI method is that it has a slow rate of adaptation although the rate is faster than adaptive loops [6,7]. By deriving an approximate probability density of the output signal-to-noise ratio (SNR) for the PCI method, we show that that if the true underlying noise covariance matrix is approximately low rank, the PCI method has a much faster rate of adaptation than the SMI method. The remainder of the paper is thus organized as follows: (1) The SMI and PCI methods of adaptive detection are reviewed, (2) an approximate probability density function of the output SNR for the PCI method is derived, SMI and PCI are related to linear regression and simple rules of thumb for comparing the performance of the PCI and SMI methods are given, (3) the results are verified experimently for a scenario using a line array to detect a signal in the presence of two near-field interferers.

1.1 SMI method of Adaptive Detection with Unknown Covariance Matrix

Let us assume that K independent multivariate zero-mean vector observations, here after called "snapshots", $\{\mathbf{d}_k\}$ of signal free noise are observed. An important rapidly adaptive approach (Reed, Mallet, Brennan [2]) for getting an estimate of the weight vector $\mathbf{w}$ is to first use the observed interference data $\{\mathbf{d}_k\}$ to get the maximum likelihood estimate of the noise covariance matrix

$$\tilde{R} = \frac{1}{K}\sum_{k=1}^{K} \mathbf{d}_k \mathbf{d}_k^H \tag{6}$$

and then use this estimate (6) in place of the true noise covariance matrix in formula (5). Thus an estimate of the weight vector is

$$\tilde{\mathbf{w}} = \tilde{R}^{-1}\mathbf{s} \quad . \tag{7}$$

This method for estimating of the weight vector (7) is called the Sample Matrix Inverse (SMI) method.

The SMI method provided more rapid adaptation than the earlier adaptive control method for determining the weight vector [6,7]. Our PCI method is intended to improve the SMI method by providing significantly more rapid adaptation under commonly occuring conditions.

Following Reed, Mallet, Brennan [2] , the effectiveness of the SMI estimate of $\mathbf{w}$ (formula (7)) can be measured using a normalized output SNR conditioned on knowing $\tilde{\mathbf{w}}$. This normalized output SNR is given by (formula (16), [2])

$$S\tilde{N}R = \frac{E[\left|\tilde{\mathbf{w}}^H e^{j\Theta}\mathbf{s}\right|^2 |\tilde{\mathbf{w}}]}{\mathbf{s}^H R^{-1}\mathbf{s} \;\; E[|\tilde{\mathbf{w}}^H \mathbf{d}|^2 |\tilde{\mathbf{w}}]} = \frac{\tilde{\mathbf{w}}^H \mathbf{s}\mathbf{s}^H \tilde{\mathbf{w}}}{\mathbf{s}^H R^{-1}\mathbf{s} \;\; \tilde{\mathbf{w}}^H R \tilde{\mathbf{w}}} \tag{8}$$

This ratio is always between zero and one

$$0 \leq S\tilde{N}R \leq 1 \quad . \tag{9}$$

Note that the quantity $\mathbf{s}^H R^{-1}\mathbf{s}$ in formula (8) is the maximum attainable output SNR [2].

The measured output SNR (8) conditioned on knowing $\tilde{\mathbf{w}}$ is a random variable because the estimated weight vector $\tilde{\mathbf{w}}$ is a function of the observed data. The quantity $S\tilde{N}R$ fully characterizes the performance of the detector based on $\tilde{\mathbf{w}}$. Brennan and Reed [3] have shown that for a fixed probability of false alarm, the probability of detection is a monotonic function of the output SNR.

Reed, Mallet, Brennan [2] and Hanumara [5] showed that for the SMI weight vector estimate, $S\tilde{N}R$ has a beta probability density of the form

$$p(\gamma) = \frac{K!}{(n-2)!(K+1-n)!}(1-\gamma)^{n-2}\gamma^{K+1-n} \quad ,0 \leq \gamma \leq 1 \ . \tag{10}$$

The expected value of $S\tilde{N}R$ is [2]

$$E[S\tilde{N}R] = \frac{K+2-n}{K+1} \quad . \tag{11}$$

A remarkable property of the SMI method is that the probability density function of the output SNR [2] does not depend on the structure of the true underlying noise covariance matrix. It is only dependent on the data vector length and the number of snapshots used to estimate the noise covariance matrix [2]. A drawback of the SMI method is that if the sample size K is small with respect to the covariance matrix order n, the performance is likely to be poor. This follows from formula (11) since if $K \approx n$ and n is large, then $E[S\tilde{N}R] << 1$.

1.2 PCI method of Adaptive Detection for Unknown, but Approximately Low Rank Covariance Matrix

In previous papers (Kirsteins and Tufts [8,9], Vaccaro et al. [10], Tufts et al. [11]), we considered the case of estimating the weight vector when the noise covariance matrix is approximately low rank and only a limited amount of noise data is available. The improved estimate of the weight vector that we advocated in the above work is

$$\tilde{\mathbf{w}} = \mathbf{s} - \sum_{k=1}^{r}(\tilde{\mathbf{v}}_k^H \mathbf{s})\tilde{\mathbf{v}}_k \tag{12}$$

where $\{\tilde{\mathbf{v}}_k\}$ are the r principal eigenvectors of the sample covariance matrix $\tilde{R}$. The above method of estimating the weight vector (12) is refered to as the Principal Component Inverse (PCI) method. We demonstrated experimently and theoretically for the near rank-1 case [8,9] that, for small number of noise snapshots, the PCI weight vector estimate will give much higher output SNR than the SMI estimate. In early 1989 we became aware of earlier, related work by Bühring [4], who advocated an approach similar to PCI in order to improve adaptive loops. No performance analysis was provided by Bühring. These results suggest that the use of prior information about the noise structure, such as rank, when estimating the weight vector can potentially improve the accuracy of the weight vector estimate. It is important to note that this "prior" information can be obtained from the data itself using conventional beamforming and filtering method before using PCI on the same data.

For a review on the optimality and other properties of eigenvalue-eigenvector decomposition based approaches in array processing, the reader is refered to Bienvenu et al. [21].

The structure of the PCI method weight vector estimate can be justified using some results of Claus et al. [16] and Liu and Nolte [17] on optimum processors. Claus et al. [16] showed in their analysis that the optimum processor removes an estimate of the interference from the data. Those eigencomponents corresponding to large eigenvalues of the noise covariance matrix R are scaled more in that removal process. This can be seen by re-writing the test statistic (formula (3)) using an identity of Claus et al. [16] as

$$\left|\mathbf{s}^H R^{-1}\mathbf{x}\right| = \sigma^{-2}\left|\mathbf{s}^H (I - P)\mathbf{x}\right| \tag{13}$$

where

$$P = \sum_{k=1}^{r} \frac{\lambda_k}{\lambda_k + \sigma^2}(\mathbf{v}_k\mathbf{v}_k^H) \quad . \tag{14}$$

The $\{\lambda_k\}$ and $\{\mathbf{v}_k\}$ in formula (14) are the r principal eigenvalues and eigenvectors of R respectively and σ^2 is the variance of the background white noise component. When the interference is strongly low rank, that is, $\lambda_k >> \sigma^2$, then

$$P \approx \sum_{k=1}^{r} \mathbf{v}_k\mathbf{v}_k^H \tag{15}$$

and the matrix $I - P$ in formula (13) corresponds to a projection operator onto the orthogonal complement of the interference space. In that case, the weight vector $(I - P)\mathbf{s}$ can be considered a generalized null steerer, ie., nulls are steered in the direction of the incoming strong interference. A similar observation was made by Gupta et al. [12] in their analysis of beam patterns when the interference consists of a small number of strong interferers plus a weak white noise component.

In addition, Liu and Nolte [17] showed for the closely related problem of detecting a rank-1 gaussian signal in strong rank-1 gaussian interference plus white gaussian noise, that the generalized null-steerer has almost the same performance as the optimum receiver.

2.0) Derivation of an Approximate Probability Density for the Output SNR of the PCI Method

We show below that when the true underlying noise covariance matrix is strongly low rank, the output SNR for the PCI method has approximately a beta probability density.

Our analysis is based on the observation that the PCI weight vector can be gotten by solving a modified linear regression problem. This in turn allows us to develop a simple linear regression based approximation for the weight vector that leads directly to the beta distribution for the output SNR.

We now briefly outline the steps used to derive the approximate probability density for the normalized output SNR of the PCI method.

1) The interpretation of detection as an estimation problem by Picinbono [13] is used to develop the linear regression realizations of SMI and PCI. In essence, the idea is that the structure of the optimum detector is the same as that for the problem where we want to find the least-squared-error linear estimator of the noise in the signal subspace from the noise components in the orthogonal complement of the signal subspace. This then directly leads to the linear regression implementations of SMI and PCI.

2) The linear regression implementation of PCI results in a very simple approximation for the PCI weight vector. This approximation is obtained from the solution to a reduced dimension linear problem. An interesting interpretation of the analysis for the case of a array is that, roughly, the PCI method is equivalent to a procedure where (a) the data is pre-filtered with a reduced set of beams that only encompasses the sector of interest (signal and interference subspace), and (b) then the SMI method is applied to the reduced set of beam outputs to estimate the weight vector.

3) The above observations lead directly to a beta probability density for the approximate normalized output SNR of the PCI method. This density is of the form

$$p(\gamma) = \frac{K!}{(K-r)!(r-1)!}\gamma^{K-r}(1-\gamma)^{r-1} \quad 0 \leq \gamma \leq 1 \tag{16}$$

where r is the interference rank and K is the number of snapshots used to estimate the noise covariance matrix.

2.1) Estimation Interpretation of Detection

The linear regression realization of SMI and PCI methods can be obtained by interpreting the optimum detection procedure as an estimation problem. The estimation interpretation of detection was pointed out by Claus et al. [16] and Picinbono [13]. Picinbono [13] showed that the structure of the optimum detector for a deterministic signal in gaussian noise is the same as that of the optimum estimator of the component of the noise in the signal subspace from the noise component in the orthogonal complement of the signal subspace. This observation was based on the fact that the likelihood-ratio test statistic can be written as [13]

$$\alpha = \left| z_0 - \mathbf{c}_{opt}^H \begin{bmatrix} z_1 \\ z_2 \\ \cdot \\ \cdot \\ \cdot \\ z_{n-1} \end{bmatrix} \right| \tag{17}$$

$$= \left| z_o - \mathbf{c}_{opt}^H \mathbf{n} \right| \tag{18}$$

where the scalars $\{z_k\}$ come from the expansion of the observed data $\mathbf{d}$ in a signal-based orthonormal basis, that is,

$$\mathbf{d} = z_0 \mathbf{s} + \sum_{k=1}^{n-1} z_k \mathbf{q}_k \tag{19}$$

where the vectors $\{\mathbf{q}_k\}$ generate the orthogonal complement of $\mathbf{s}$. Later, in section **2.3** we will make a particular choice of basis vectors $\{\mathbf{q}_k\}$. The vector $\mathbf{c}_{opt}$ is given by

$$\mathbf{c}_{opt} = \Gamma_2^{-1} \Gamma_{21} \tag{20}$$

where

$$E[\mathbf{z}\mathbf{z}^H] = \Gamma = \begin{bmatrix} \Gamma_1 & \Gamma_{12} \\ \Gamma_{21} & \Gamma_2 \end{bmatrix} \tag{21}$$

and $\mathbf{z} = [z_0 z_1 \ldots z_{n-1}]^T$. Γ_1 is a scalar, Γ_{12}, Γ_{21} are row and column vectors of length $n-1$, and Γ_2 is a $n-1 \times n-1$ matrix. The matrix Γ is just the covariance matrix of the noise in the signal-based coordinate system. By inspection from the likelihood-ratio test statistic (formula (18)), the optimum weight vector in the new coordinate system is

$$\mathbf{w}_o = \begin{bmatrix} 1 \\ \mathbf{c}_{opt} \end{bmatrix} . \tag{22}$$

This scaled version of the optimum weight vector of formula (5) (evaluated in the signal-based coordinate system) preserves the signal to noise ratio. Note that the scalar z_0 represents the component of the noise in the signal subspace and

$z_1, z_2, \ldots, z_{n-1}$ represent the components of the noise in the orthogonal complement of the signal subspace. The connection of detection to estimation was made in the following manner: Picinbono [13] pointed out that that the scalar $\mathbf{c}_{opt}^H \mathbf{n}$ is the best linear mean-square estimate of z_0 from $\mathbf{n}$. Thus, another interpretation of (18) is that the optimum detector attempts to estimate and then remove the component of the interference in the signal subspace using the noise component in the orthogonal complement of the signal subspace.

2.2) Linear Regression Realization of SMI and PCI

SMI Method

The above interpretations and observations suggest that it should be possible to reformulate SMI and PCI methods as linear regression procedures where we attempt to estimate the component of the noise in the signal subspace. An obvious approach for getting an estimate of $\mathbf{c}_{opt}$ is to solve

$$\min_{\mathbf{c}} \|Y^H \mathbf{c} + x^H\|_F \tag{23}$$

where

$$x = \begin{bmatrix} \mathbf{s}^H \mathbf{d}_1 & \mathbf{s}^H \mathbf{d}_2 & \cdots & \mathbf{s}^H \mathbf{d}_K \end{bmatrix} \tag{24}$$

and

$$Y = \begin{bmatrix} \mathbf{q}_1^H \mathbf{d}_1 & \mathbf{q}_1^H \mathbf{d}_2 & \cdots & \mathbf{q}_1^H \mathbf{d}_K \\ \mathbf{q}_2^H \mathbf{d}_1 & \mathbf{q}_2^H \mathbf{d}_2 & \cdots & \mathbf{q}_2^H \mathbf{d}_K \\ \cdot & \cdot & \cdots & \cdot \\ \cdot & \cdot & \cdots & \cdot \\ \cdot & \cdot & \cdots & \cdot \\ \mathbf{q}_{n-1}^H \mathbf{d}_1 & \mathbf{q}_{n-1}^H \mathbf{d}_2 & \cdots & \mathbf{q}_{n-1}^H \mathbf{d}_K \end{bmatrix} \tag{25}$$

Recall that $\{\mathbf{d}_k\}$ is the observed interference data with no signal present. The row vector x represents the observed noise components in the signal subspace and the matrix Y represents the observed noise component in the orthogonal complement of the signal subspace. From formula (22) it follows that an estimate of the weight vector is

$$\tilde{\mathbf{w}}_o = \begin{bmatrix} 1 \\ \tilde{\mathbf{c}}_{opt} \end{bmatrix} . \tag{26}$$

where $\tilde{c}_{opt}$ is the solution to (23). In fact, the above estimate (formula (26)) is an SMI method weight vector estimate. The details of the proof are provided in Appendix A.

PCI Method

The linear regression realization of SMI suggests that it should be possible to formulate the PCI method as some type of linear regression problem. A procedure analogous to PCI is to solve (23) using only the low rank component of the noise, ie., null out just the low rank part of noise in the signal subspace. We can get an estimate of the low rank part of the noise in the signal subspace and its' orthogonal complement by finding a low rank approximation to the augmented matrix $Z^H = [x^H | Y^H]$. Recall from formulas (24) and (25) that the row vector x represents the observed noise components in the signal subspace and the matrix Y represents the observed noise components in the othogonal complement of the signal subspace. The best rank-r approximation (in the least-squares sense) to Z^H in terms of the singular-value-decomposition of Z^H [14] is

$$Z_r^H = \sum_{k=1}^{r} \tilde{\sigma}_k \tilde{\mathbf{u}}_k \tilde{\mathbf{v}}_k^H \tag{27}$$

where $\tilde{\sigma}_k$ are the singular values and the $\tilde{\mathbf{u}}_k$ and $\tilde{\mathbf{v}}_k$ are the left and right singular vectors respectively of Z^H. Thus, a scaled version of the PCI weight vector of formula (12), written in the signal-based coordinate system which was used in formula (19) is (see Appendix B for proof)

$$\tilde{\mathbf{w}}_{PCI} = \begin{bmatrix} 1 \\ \tilde{\mathbf{c}}_{PCI} \end{bmatrix} \tag{28}$$

where $\tilde{\mathbf{c}}_{PCI}$ is the minimum-norm solution to

$$\min_{\mathbf{c}} \; \|Y_r^H \mathbf{c} + x_r^H\|_F \tag{29}$$

and Y_r^H and x_r^H are obtained from the partitioned matrix

$$Z_L^H = [x_r^H | Y_r^H] \tag{30}$$

where, recall, Z_r^H is the best rank-r approximation to the interference plus white noise data matrix Z^H in the least-squares sense. The reason for choosing the minimum-norm solution above is to minimize the output white noise power after removing the low rank component of the interference. Our next step is to get an approximate solution to (29) in order to obtain an approximation to $\tilde{\mathbf{w}}_{PCI}$.

2.3) Approximating the PCI Method Weight Vector Estimate

Let us begin by noting that $\tilde{c}_{PCI}$ can also be gotten from the solution to

$$\min_{\mathbf{c}} \|Y_r^H \mathbf{c} + x^H\|_F \tag{31}$$

since

$$\|Y_r^H \mathbf{c} + x^H\|_F = \|Y_r^H \mathbf{c} + x_r^H\|_F + \|x^H - x_r^H\|_F \quad . \tag{32}$$

We now propose another estimator of $\mathbf{c}_{opt}$ that is similar in structure and spirit to PCI, but turns out to be easier to analyse. The proposed estimator is

$$\min_{\mathbf{c}} \|\tilde{Y}_r^H \mathbf{c} + x^H\|_F \tag{33}$$

where $\tilde{Y}_r^H$ is the best rank-r approximation (in the least-squares sense) to Y^H. Note that the two estimators (formulas (31) and (33)) differ in only that Y_r^H is gotten from the rank-r approximation to the augmented matrix $[x^H|Y^H]$ while $\tilde{Y}_r^H$ is the best rank-r approximation to Y^H. We conjecture that when the interference is strongly rank-r and not too close to the signal, then $Y_r^H \approx \tilde{Y}_r^H$. The reasoning is as follows: If the signal is approximately orthogonal to the interference, then most of the low rank part of interference appears in Y^H and only little appears in x^H. It follows that augmenting Y^H with x^H should not significantly improve the estimation accuracy of the low rank component of the noise present in Y^H. Therefore, the two estimators should be roughly the same.

Approximating $\tilde{Y}_r^H$

The next step in the analysis is obtain an approximation to $\tilde{Y}_r^H$. We begin by noting that the analysis can be simplied by choosing an appropriate set of $\mathbf{q}_k$'s. Recall that we have not yet specified any particular choice for the $\mathbf{q}_k$'s in formula (19). Any orthonormal set of $\mathbf{q}_k$'s that generates the orthogonal complement of the signal subspace can be used. A reasonable choice for the $\mathbf{q}_k$'s is one that satisfies

$$E[\overline{(\mathbf{q}_k^H \mathbf{d})}(\mathbf{q}_j^H \mathbf{d})] = 0 \quad , \quad k \neq j \tag{34}$$

$$\mathbf{q}_k^H R \mathbf{q}_j = 0 \quad , \quad k \neq j \tag{35}$$

ie. the random variables $\mathbf{q}_k^H \mathbf{d}$ and $\mathbf{q}_j^H \mathbf{d}$ are uncorrelated if $k \neq j$. It can be shown that a set of $\mathbf{q}_k$'s that satisfy (35) are the $n-1$ principal eigenvectors of the matrix $P_\perp R P_\perp$ where $P_\perp = I - \mathbf{s}\mathbf{s}^H$. We note that $P_\perp R P_\perp$ can be interpreted as the covariance matrix of the noise in the orthogonal complement of the signal subspace.

With the $\mathbf{q}_k$'s constructed using the above procedure, the low rank component of the noise appears only in the first r columns of Y^H. This suggests that using these basis vectors, an approximation to $\tilde{Y}_r^H$ is

$$\tilde{Y}_r^H \approx \left[\ \tilde{Y}_o^H \ \ |0\ \right] \tag{36}$$

where

$$\tilde{Y}_o^H = \begin{bmatrix} \mathbf{d}_1^H\mathbf{q}_1 & \mathbf{d}_1^H\mathbf{q}_2 & \cdots & \mathbf{d}_1^H\mathbf{q}_r \\ \mathbf{d}_2^H\mathbf{q}_1 & \mathbf{d}_2^H\mathbf{q}_2 & \cdots & \mathbf{d}_2^H\mathbf{q}_r \\ \cdot & \cdot & \cdots & \cdot \\ \cdot & \cdot & \cdots & \cdot \\ \cdot & \cdot & \cdots & \cdot \\ \mathbf{d}_K^H\mathbf{q}_1 & \mathbf{d}_K^H\mathbf{q}_2 & \cdots & \mathbf{d}_K^H\mathbf{q}_r \end{bmatrix} . \tag{37}$$

Weight Vector Approximation

Substituting (36) in place of Y_r^H in formula (33)

$$\min_{c} \|[\tilde{Y}_o^H|0]\mathbf{c} + x^H\|_F \tag{38}$$

and then finding the minimum-norm solution to (38) leads us to the desired approximation to $\tilde{\mathbf{c}}_{opt}$. The minimum-norm solution to (38) is

$$\mathbf{c}_I = \begin{bmatrix} \tilde{\mathbf{c}}_r \\ 0 \end{bmatrix} = \begin{bmatrix} -(\tilde{Y}_o\tilde{Y}_o^H)^{-1}\tilde{Y}_o x^H \\ 0 \end{bmatrix} . \tag{39}$$

Thus, an approximation to $\tilde{\mathbf{w}}_{PCI}$ of formula (28), when the special basis vectors of formula (35) are used, is

$$\tilde{\mathbf{w}}_{PCI} \approx \begin{bmatrix} 1 \\ \tilde{\mathbf{c}}_r \\ 0 \end{bmatrix} . \tag{40}$$

2.4) Approximate Probability Density for the SNR of the PCI Method

It is now shown by analogy to the SMI method that the output SNR of PCI method has approximately a beta distribution. We first note that, because of the scaling of the PCI weight vector introduced in formulas (28)) and (40), the output signal power is normalized to unity. The output normalized SNR of the PCI weight vector approximation (formula (40)) in the signal-based coordinate system is

$$SNR = \frac{1}{(\mathbf{s}_o^H\Gamma^{-1}\mathbf{s}_o)[1|\tilde{\mathbf{c}}_r^H|0]\Gamma[1|\tilde{\mathbf{c}}_r^H|0]^H} \tag{41}$$

where $\mathbf{s}_o = [10\ldots0]^T$. Next note that Γ can be partitioned as (assuming the $\mathbf{q}_k$'s are selected according to previous criterion)

$$\Gamma = \begin{bmatrix} \Gamma_0 & 0 \\ 0 & \Lambda_0 \end{bmatrix} \tag{42}$$

where Γ_0 is a $r+1\times r+1$ matrix and Λ_o is a and $n-r-1\times n-r-1$ diagonal matrix. Plugging the partitioned form of Γ in (41), and after some algebraic manipulation,

SNR can be re-written as

$$SNR = \frac{1}{(\mathbf{s}_r^H \Gamma_0^{-1} \mathbf{s}_r)[1|\tilde{c}_r^H]\Gamma_0[1|\tilde{c}_r^H]^H} \tag{43}$$

where $\mathbf{s}_r = [10\ldots0]^T$ is of dimension $r+1$. We can now make the following observation: Comparing the structures of (41) and (43), we conclude that formula (43) also represents the output SNR for a $(r+1)$-dimension problem, where Γ_0 is the covariance matrix of the vector $\mathbf{z}_r = [z_0 z_1 \ldots z_r]^T$ (see section **2.1**). Recall that $\tilde{c}_r$ is obtained as the solution to

$$\min_{\mathbf{c}} \|\tilde{Y}_o^H \mathbf{c} + x^H\|_F \tag{44}$$

in which $\tilde{Y}_o$ is defined by formula (37). It can be easily shown that the solution of the above problem (44) is the SMI estimate of $\mathbf{c}_{opt}$ for the reduced dimension set of data $\mathbf{z}_r$ using K snapshots of interference data (follows the proof in Appendix A except Y is replaced with $\tilde{Y}_o$). From the above observations, we now infer that the distribution of (41) is the same as for the SMI method weight vector estimate, except the dimensionality is reduce to $r+1$.

Thus, it follows that the output SNR for the PCI method has approximately a beta probability density. This beta probability density function is of the form (letting $\gamma = SNR_{PCI}$)

$$p(\gamma) = \frac{K!}{(K-r)!(r-1)!}\gamma^{K-r}(1-\gamma)^{r-1} \quad ,0 < \gamma < 1 \tag{45}$$

where r is the interference rank and K is the number of snapshots of interference data data used to estimate R. It is remarkable that when the interference is strongly low rank, the performance of the PCI method is essentially independent of the structure of the true, low rank, noise covariance matrix.

<u>Discussion</u>

We can now make an interesting interpretation of the PCI method. Let us consider a case where a array is used to detect a signal. Then, the above results suggest that the PCI method is roughly equivalent to a procedure based on first forming $r+1$ "beams" using the vectors $\{\mathbf{s}, \mathbf{q}_1, \ldots, \mathbf{q}_r\}$ and then applying the SMI method on that reduced set of beam outputs. In essence, the improvement of the PCI method stems from performing a pre-filtering of data where we limit the processing to only the sector of interest (signal and interference subspace), rather than the entire space.

It is important to note that some of this pre-filtering can be carried out in a prescribed way, ie., independent of the currently observed data. That is, the observed data for the PCI can be roughly filtered (spatially and temporally) data, rather than direct sensor samples. Adams et al. [18] suggested a similar approach for an adaptive nulling procedure where SMI was applied on beam output data rather than direct sensor data. Essentially, the idea was to steer a small number of beams in the direction of the signal and interference, and then apply SMI on the beam outputs directly [18].

2.5) Signal-Corrupted Noise Data

We now note that our results can be extended to analyse the PCI method for the case when the observed noise data is corrupted by signal. Boroson [15] did an analysis of the SMI method when the noise is corrupted by signal and provided a formula for the distribution of SNR. Specifically, Boroson [15] considered the case when the observed data is of the form

$$\mathbf{x} = \mathbf{d} + \beta \mathbf{s} \tag{46}$$

where β is a zero-mean complex gaussian distributed random variable with variance σ_s^2 and, recall, d is the interference.

Provided that the contaminating signal is weak, our assumptions and approximations used in deriving the approximate SNR are still valid. Thus, the realization of PCI as a reduced dimension SMI implementation is still true and the results of Boroson [15] are directly applicable. From Boroson [15], it can be shown that an approximate probability density of SNR for the PCI method in the signal corrupted case is

$$p_s(\gamma) = p\left(\frac{\gamma(1+T_o)}{1+\gamma T_o}\right)\left[\frac{1+T_o}{1+\gamma T_o} - \frac{\gamma T_o(1+T_o)}{(1+\gamma T_o)^2}\right] \tag{47}$$

where $p(\gamma)$ is the same as in formula (45) and $T_o = \sigma_s^2[10\ldots0]\Gamma_0[10\ldots0]^T$.

2.6) Simple Rules for Comparing Performance of PCI and SMI Methods

The performance of the PCI and SMI methods can now be quantitatively compared and analysed using measures derived from their respective probability density functions of the output SNR. Note that since the output SNR for the PCI method has also a beta distribution, the same measures and rules developed for the SMI method are applicable to the PCI method (adjusting the data dimension to $r+1$).

A useful measure suggested by Reed, Mallet, Brennan [2] is to consider the snapshots of noise data K needed to maintain the expected value of the output SNR to greater than one-half, ie. within 3 dB of the optimum. They showed that at least $K_{SMI} = 2n - 3$ snapshots are needed by the SMI method. Adjusting the data dimension in the above formula to $r+1$, the number of snapshots needed by the PCI method is $K_{PCI} = 2(r+1) - 3$.

In the work of Boroson [15] on sample size considerations for adaptive arrays, it was pointed out in a discussion on the SMI method that the moments of the output SNR do not always necessarily provide an accurate indication of performance.

Boroson [15] suggested that the performance of the SMI method could be more accurately quantified by considering the probability of obtaining a "bad" estimate of the weight vector, ie., one whose output SNR falls below some desired level, say δ. Boroson [15] provided the following formula to calculate $P(SNR < \delta)$ for the SMI method:

$$P_{SMI}(SNR < \delta) = \sum_{m=0}^{n-2} \binom{K}{m} (1-\delta)^m \delta^{K-m} \quad . \tag{48}$$

Since the output SNR for the PCI method has approximately a beta distribution, the corresponding formula for PCI can be written by adjusting the data dimension to $r+1$ as follows:

$$P_{PCI}(SNR < \delta) = \sum_{m=0}^{r-1} \binom{K}{m} (1-\delta)^m \delta^{K-m} \quad . \tag{49}$$

For example, it was shown by Boroson that when $n > 3$ and $r+1 > 3$ (adjusting the dimensions for the PCI method), $K \geq 4n$ ensures that $P_{SMI}(SNR < 1/2) < .0032$ and $K \geq 4(r+1)$ ensures that $P_{PCI}(SNR < 1/2) < .0032$.

From the these formulas, it can be seen that if n is large and $n >> r$, the SMI method needs much more data than the PCI method to attain the above desired performance criteria.

3.0 Experimental Results

In this section some examples are presented to verify the accuracy of the approximate SNR derived for the PCI method (see section **2.4**) for both the noise only and signal-corrupted data case and then show the potential improvement of the PCI method over the conventional SMI method of estimating the weight vector.

Before we begin, let us briefly summarize the main approximations used in the analysis. They are:

I) The exact PCI method weight vector estimate (in the signal-based coordinate system) is gotten from the unique minimum-norm solution of the problem

$$\min_{\mathbf{c}} \|Y_r^H \mathbf{c} + x^H\|_F \quad (augmented) \tag{50}$$

This solution is replaced by the unique minimum-norm solution of the problem

$$\min_{\mathbf{c}} \|\tilde{Y}_r \mathbf{c} + x^H\|_F \quad (non-augmented) \tag{51}$$

where, recall, Y_r^H is obtained from the rank-r approximation to the augmented matrix $[x^H|Y^H]$ and $\tilde{Y}_r^H$ is the rank-r approximation to Y^H.

II) The minimum-norm solution of the problem of expression (51) is further replaced by an approximate solution as described in section **2.3**. Substitution of the resulting approximate PCI weight vector into the normalized SNR (formula (8)) gives us the desired approximation (see section **2.4**)

$$SNR = \frac{1}{(\mathbf{s}_r^H \Gamma_0^{-1} \mathbf{s}_r)[1|\tilde{\mathbf{c}}_r^H]\Gamma_0[1|\tilde{\mathbf{c}}_r^H]^H} \tag{52}$$

to the PCI normalized output SNR, where, recall, $\tilde{\mathbf{c}}_r = -(\tilde{Y}_o \tilde{Y}_o^H)^{-1} \tilde{Y}_o x^H$ (formula (39)).

Experimental Scenario

We consider a scenario where a twenty sensor equi-spaced line array is used to detect a weak monochromatic plane wave signal arriving at broadside in the presence of interference from two nearby strong gaussian interferers of equal strength (unit variance) and a weak gaussian isotropic component. The distance between adjacent sensors is one half the signal wavelength, denoted as λ. The near field interferers are modeled as monochromatic point sources (wavelength λ) with Rayleigh distributed amplitudes and uniformly distributed phases, say due to channel fading or fluctuating reflectors. The weak isotropic component is assumed to be spatially and temporally uncorrelated. The interference-to-noise ratio (INR) is defined as $INR = 10 log_{10} \frac{1}{20\sigma_n^2}$ where σ_n^2 is the variance of the isotropic component.

Twenty independent snapshots of noise-only array data at an INR of 10 dB are used for adaptation in our experimental results. One point source interferer was placed at *range* $= 57\lambda$, *bearing* $= 84.26$ *degrees* and the other at *range* $=$ 57λ, *bearing* $= 95.74$ *degrees* (a range of 57λ corresponds to six array-lengths). The range and bearing of the point source interfererers are measured with respect to the center and orientation of the line array respectively. The bearings are chosen so that the signal would be separated from both interferers by a beam-width if the interference wavefronts were planar. Note that the range of the interferers are sufficiently close to result in significant interference wavefront curvature at the array.

3.1 Testing the Accuracy of the Approximate Output SNR for the PCI Method through Simulation

The accuracy of the approximations used in the analysis are judged by how closely the output SNR based on these approximations follows that of the PCI method, since only the resultant output SNR is of interest here. Briefly, our approach is as follows: In the simulations, each new realization of twenty noise snapshot vectors produces new weight vectors, both approximate PCI and exact PCI (formula (12)). The normalized output SNR of these new weight vectors are calculated (using formula (8) in the appropriate coordinate system) and then tabulated on a trial-by-trial basis for the simulation run. We then use two different methods to judge the accuracy of the approximate PCI output SNR. First, the exact PCI and approximate PCI output SNR's are visually compared on a trial-by-trial basis using a scattergram. Secondly, an estimate of the exact PCI output SNR probability density is obtained by constructing a histogram using the experimently observed data and then compared against the theoretically derived probability density function.

3.1.0 Signal-Free Noise Data

Comparison of PCI Against Non-Augmented Method

The true PCI output SNR is compared with the output SNR of the weight vector estimate gotten from the non-augmented solution (formula (33)) for each separate set of realizations of the noise vectors used for adaptation. These pairs of PCI and non-augmented solution weight vector estimate SNR values are shown in the scattergram of Fig. 1 for 300 sets of 20 noise observations. The results from Fig. 1 indicate that the output SNR of the non-augmented method is a reasonable approximation to the output SNR of the PCI method.

Comparison of PCI Against the Approximate Output SNR

The accuracy of the derived approximate output SNR was tested by comparing the approximate output SNR (formula (52)) against the true PCI output SNR for each separate set of realizations of the noise vectors used for adaptation. These pairs of approximate and exact SNR values are shown in the scattergram of Fig. 2 for 300 sets of 20 noise observations. In Fig. 3 the experimental histogram of the exact SNR random variable, using 300 trial sets, is compared with the one standard-deviation confidence curves which are obtained from the derived, approximate probability density of SNR. The results from Fig. 2 and Fig. 3 indicate that the SNR approximations and derived probability density function are accurate for this example of interference and signal.

3.1.1 Signal-Corrupted Data

The above experiments are repeated for the case when the observed interference data is corrupted by signal. We use the same signal contamination model as was given in section **2.5**. Recall that the received signal-corrupted data is

$$\mathbf{x} = \mathbf{d} + \beta \mathbf{s} \tag{53}$$

where β is a zero-mean complex gaussian random variable with variance σ_s^2. For our experimental results, we used an signal-to-interference ratio (SIR) of -15db where $SIR = 10log_{10}\sigma_s^2$. The scattergrams (see Fig. 4, 5) and histogram (see Fig. 6) indicate that the approximations and probability density function are accurate even for the signal-corrupted data case.

3.2 Evaluation of the SNR Probability Density Functions for the PCI and SMI Methods

Signal-Free Noise Data

The approximate probability density (formula (45)) for the PCI method weight vector estimate is plotted (see Fig. 7) for the interference scenario defined earlier. For comparison we also plot the probability density function of the normalized output SNR for the SMI method (formula (10)). It can be shown that (see Fig. 7), for the same amount of data used for adaptation, the PCI method provides significantly greater values of SNR with high probability.

Signal-Corrupted Noise Data

Next we evaluate the PCI (formula (47)) and SMI [15] probability density functions of the output SNR for the case when the noise data is corrupted by signal at SIR=-15 dB. Both densities are plotted in Fig. 8. The plots indicate that for low levels of signal contamination, the PCI method still performs reasonably well and is superior to SMI..

Conclusion

We have derived an approximate probability density function of the output SNR for the PCI method that is accurate over a reasonable range of noise structures and signal types. In addition we have provided useful formulas that quantitatively analyze the performance of PCI method in terms of the number of noise snapshots needed to attain some desired level of performance and compare it against the SMI method. Our results show that when the noise covariance matrix is strongly low rank, the PCI method has an adaptation rate that is significantly faster than the SMI method.

References

1) L.W. Brooks and I.S. Reed, "Equivalence of the Likelihood Ratio Processor, The Maximum Signal-to-Noise Ratio Filter, and the Wiener Filter," IEEE Trans. Aerospace and Electronic Systems, pp. 690-692, Sept. 1972.

2) L.S. Reed, J.D. Mallett and L.E. Brennan, "Rapid Convergence Rate in Adaptive Arrays," IEEE Trans. Aerospace and Electronic Systems, Vol. AES-10, No. 6, pp. 853-863, Nov. 1974.

3) L.E. Brennan and L.S. Reed, "Theory of Adaptive Radar," IEEE Trans. Aerospace and Electronic Systems, Vol. AES-9, No. 2, pp. 237-251, March 1973.

4) W. Bühring, "Adaptive Orthogonal Projection for Rapid Converging Interference Suppression," Electronics Letters, Vol. 14, No. 16, pp. 515-516, Aug. 3, 1978.

5) R.C. Hanumara, "An Alternate Derivation of the Distribution of the Conditioned Signal-to-Noise Ratio," IEEE Trans. Ant. and Prop., Vol. AP-34, No. 3, pp. 463-464, March 1986.

6) S.P. Applebaum, "Adaptive Arrays," IEEE Trans. Ant. and Prop., Vol. AP-24, No. 5, pp. 585-598, Sept. 1976.

7) G. Bienvenu and J.L. Vernet, "Enhancement of Antenna Performance by Adaptive Processing," NATO ASI, Loughborough, Aug. 1972.

8) I.P. Kirsteins and D.W. Tufts, "On the Probability Density of Signal-to-Noise Ratio in Improved Adaptive Detector," ONR Technical Report, Oct. 1984.

9) I.P. Kirsteins and D.W. Tufts, "On the Probability Density of Signal-to-Noise Ratio in Improved Adaptive Detector," Proc. ICASSP-85, pp. 572-575, March 1985, Tampa, Florida.

10) R.J. Vaccaro, D.W. Tufts, G.F. Boudreaux-Bartels, chapter in *Singular Value Decomposition and Signal Processing*, Advances in Principal Component Signal Processing, Proc. of the Int. Workshop on SVD and Signal Processing, Sept. 1987, Les Houches, France.

11) D.W. Tufts, C.D. Melissinos and I.P Kirsteins, "Very Fast Adaptive Array Processing," Proc. of 25'th IEEE Conf. on Decision and Control, pp. 923-928,

Dec. 1986, Athens, Greece.

12) I.J. Gupta and A.A. Ksienski, "Dependence of Adaptive Array Performance in Conventional Design," IEEE Trans. Ant. and Prop., Vol. AP-30, No. 4, pp. 549-553, July 1982.

13) B.C. Picinbono, "A Geometric Interpretation of Signal Detection and Estimation," IEEE Trans. Info. Theory, Vol. IT-26, No. 4, pp. 493-497, July 1980.

14) C. Ekart and G. Young, "The Approximation of One Matrix by Another of Lower Rank," Psychometrika, 1(1936), pp. 211-218.

15) D.M. Boroson, "Sample Size Considerations for Adaptive Arrays," IEEE Trans. Aerospace and Electronic Systems,
Vol. AES-16, No. 4, pp. 446-451, July 1980.

16) A.J. Claus, T.T. Kadota, and D.M. Romain, "Efficient Approximation of a Family of Noises for Application in Adaptive Spatial Processing for Signal Detection," IEEE Trans. Info. Theory, Vol. IT-26, No. 5, pp.588-595, Sept. 1980.

17) S.C. Liu and L.W. Nolte, "Performance Evaluation of Array Processors for Detecting Gaussian Signals, " IEEE Trans. Acoust., Speech, and Signal Proc., Vol. ASSP-28, No. 3, pp. 328-333, June 1980.

18) R.N. Adams, L.L. Horowitz, and K.D. Senne, "Adaptive Main-Beam Nulling for Narrow-Beam Antenna Arrays," IEEE Trans. Aero. and Electr. Systems, Vol. AES-16, No. 4, pp. 509-516, July 1980.

19) P.J. Davis, *Interpolation and Approximation*, Dover Publications, New York, 1975.

20) W.H. Beyer, *CRC Standard Mathematical Tables*, CRC Press, Boca Raton, Florida, 1980.

21) G. Bienvenu and L. Koop, "Optimality of High Resolution Array Processing Using Eigensystem Approach," IEEE Trans. ASSP, Vol. ASSP-31, No. 5, pp. 1235-1248, Oct. 1983.

Appendix A) Equivalence to the SMI Weight Vector Estimate

We want to show that the weight vector estimate (26)

$$\tilde{\mathbf{w}}_o = \begin{bmatrix} 1 \\ \tilde{\mathbf{c}}_{opt} \end{bmatrix} \tag{54}$$

$$= \begin{bmatrix} 1 \\ -(YY^H)^{-1}Yx^H \end{bmatrix} \tag{55}$$

(it can be easily shown that $\tilde{\mathbf{c}}_{opt} = -(YY^H)^{-1}Yx^H$.) is the same, except for a scale factor, as the SMI weight vector estimate (formula (7)) .

The SMI weight vector estimate in the signal-based coordinate system is

$$\tilde{\mathbf{w}}_o = \tilde{\Gamma}^{-1}\mathbf{s}_o \tag{56}$$

where $\tilde{\Gamma}$ and $\mathbf{s}_o$ represent the maximum likelihood estimate of the noise covariance matrix and signal vector ($\mathbf{s}_o = [10\ldots0]^T$) respectively in the signal-based coordinate system. Next note that $\tilde{\Gamma}$ can be written in terms of the observed data x and Y as

$$\tilde{\Gamma} = \frac{1}{K}\begin{bmatrix} x \\ Y \end{bmatrix}[x^H|Y^H] \tag{57}$$

$$= \frac{1}{K}\begin{bmatrix} xx^H & xY^H \\ Yx^H & YY^H \end{bmatrix} . \tag{58}$$

Finally, substituting (58) in place of $\tilde{\Gamma}$ in (56) and using the matrix inversion formula for partitioned matrices [20] to invert $\tilde{\Gamma}$, we get

$$\tilde{\mathbf{w}}_o = \frac{K}{(xx^H - xY^H(YY^H)^{-1}Yx^H)}\begin{bmatrix} 1 \\ -(YY^H)^{-1}Yx^H \end{bmatrix} . \tag{59}$$

By inspection of formulas (59) and (55) it can be seen that the two weight vector estimates are identical except for a scale factor.

Appendix B) Equivalence to the PCI Weight Vector Estimate

We want to show that the weight vector estimate (28)

$$\tilde{\mathbf{w}}_{PCI} = \begin{bmatrix} 1 \\ \tilde{\mathbf{c}}_{PCI} \end{bmatrix} \tag{60}$$

is the same, except for a scale factor, as the PCI weight vector estimate (formula (12)) . Recall that $\tilde{\mathbf{c}}_{PCI}$ is the minimum-norm solution to

$$\min_{\mathbf{c}} \|Y_r^H \mathbf{c} + x_r^H\|_F \tag{61}$$

where Y_r^H and $\mathbf{x}_r^H$ are gotten from the partitioned matrix $Z_r^H = [x_r^H | Y_r^H]$, recalling that Z_r^H is the best rank-r approximation to the interference plus white noise data matrix in the least-squares sense. In general we are able to find a solution to (61) that results in zero error, ie., $\|Y_r^H \tilde{\mathbf{c}}_{PCI} + x_r^H\|_F = 0$. Therefore, the $\tilde{\mathbf{w}}_{PCI}$ must lie in the complex conjugate of the orthogonal complement of the row space of Z_r^H. Thus, $\tilde{\mathbf{w}}_{PCI}$ can be expanded as

$$\tilde{\mathbf{w}}_{PCI} = \begin{bmatrix} 1 \\ \tilde{\mathbf{c}}_{PCI} \end{bmatrix} = \sum_{k=r+1}^{n} \alpha_k \tilde{\mathbf{v}}_k \tag{62}$$

where the $\tilde{\mathbf{v}}_k$ are the right singular vectors of Z^H.

The only constraint on (62) is that the first element must be one. Thus, premultiply both sides of (62) by the vector $\mathbf{s}_o^T = [10\ldots0]^T$. We then obtain the following equation that must be satisfied:

$$1 = \sum_{k=r+1}^{n} \alpha_k \mathbf{s}_o^T \tilde{\mathbf{v}}_k \quad . \tag{63}$$

The minimum-norm solution to (63) with respect to the α_k's is, using the Cauchy-Schwarz inequality [19],

$$\alpha_k = \frac{\tilde{\mathbf{v}}_k^H \mathbf{s}_o}{\sum_{k=r+1}^{n} |\mathbf{s}_o^T \tilde{\mathbf{v}}_k|^2} \quad . \tag{64}$$

Thus,

$$\tilde{\mathbf{w}}_{PCI} = \begin{bmatrix} 1 \\ \tilde{\mathbf{c}}_{PCI} \end{bmatrix} = \frac{1}{\sum_{k=r+1}^{n} |\mathbf{s}_o^T \tilde{\mathbf{v}}_k|^2} \sum_{k=r+1}^{n} (\tilde{\mathbf{v}}_k^H \mathbf{s}_o) \tilde{\mathbf{v}}_k \quad . \tag{65}$$

We now show the equivalence of (65) to the PCI weight vector estimate (formula (12))

$$\tilde{\mathbf{w}} = \mathbf{s}_o - \sum_{k=1}^{r} (\tilde{\mathbf{v}}_k^H \mathbf{s}_o) \tilde{\mathbf{v}}_k \tag{66}$$

where now the $\tilde{\mathbf{v}}_k$ are the principal eigenvectors of estimated interference covariance matrix, $\tilde{\Gamma} = \frac{1}{K} Z Z^H$, in the signal-based coordinate system. Note that the right singular vectors of Z^H are the eigenvectors of $\tilde{\Gamma}$. The equivalence to (65) can be shown by re-writing (66) as

$$\tilde{\mathbf{w}} = \sum_{k=r+1}^{n} (\tilde{\mathbf{v}}_k^H \mathbf{s}_o)\tilde{\mathbf{v}}_k \quad . \tag{67}$$

Note that formulas (65) and (67) differ only by a scale factor. Therefore, the two weight vector estimates are the same except for a scale factor.

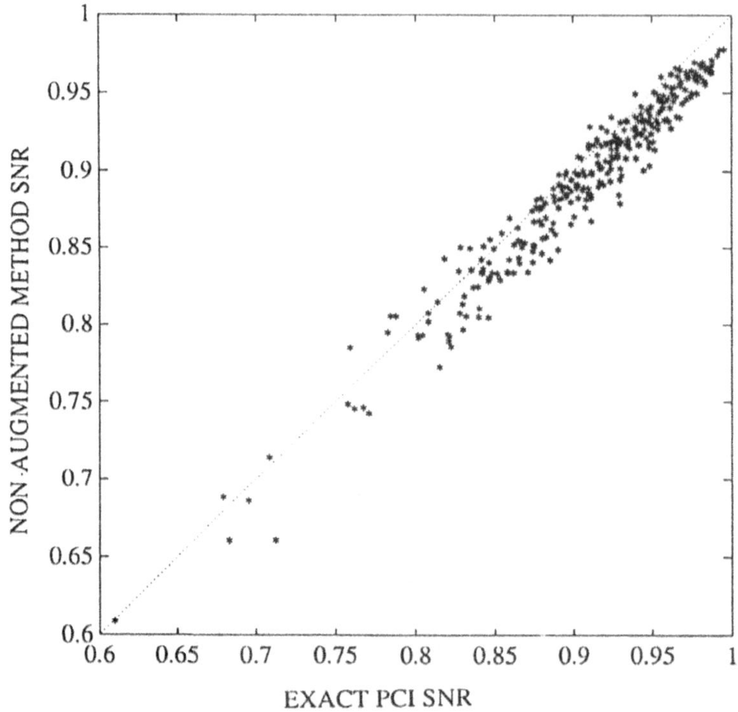

Fig. 1) **Scattergram of the non-augmented method SNR vs. exact PCI SNR based on 300 trials. The observed noise data is signal-free.**

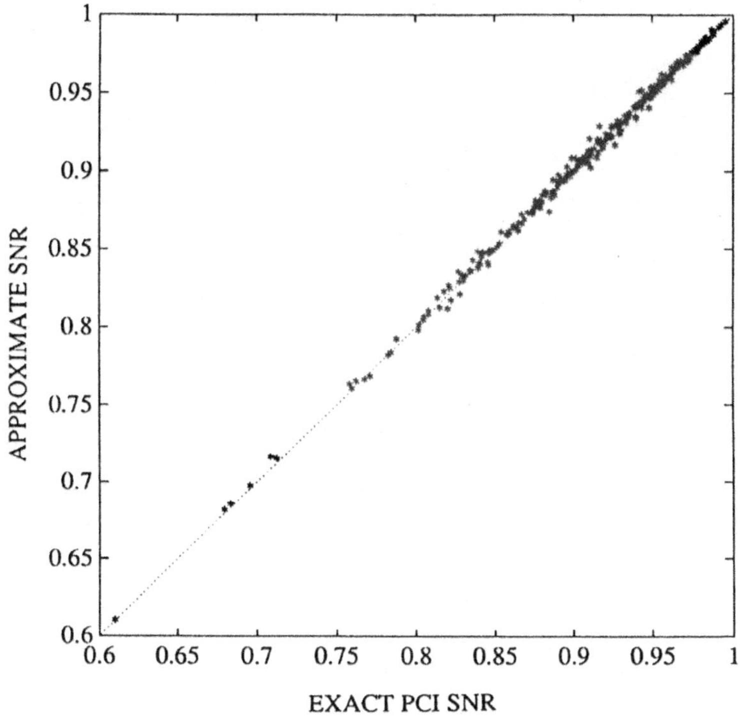

Fig. 2) Scattergram of the approximate SNR vs. exact PCI SNR based on 300 trials. The observed noise data is signal-free.

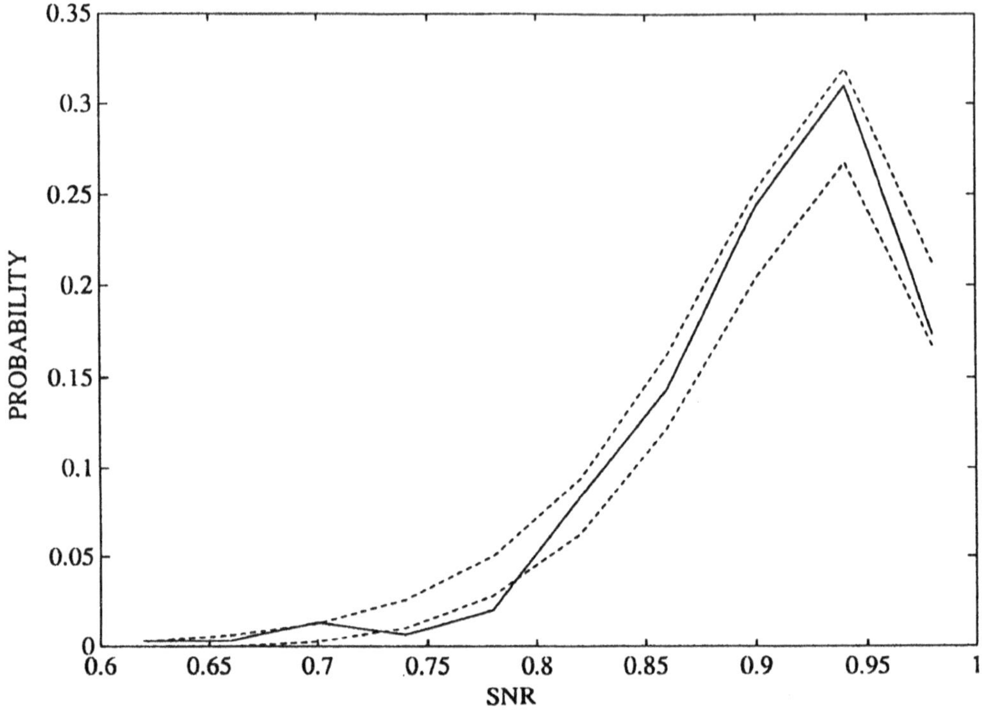

-------- CONFIDENCE BOUND FOR EACH
HISTOGRAM BIN BASED ON THE ESTIMATED
STANDARD DEVIATION ABOUT THE EXPECTED
BIN FREQUENCY USING THE BINOMIAL DISTRIBUTION

Fig. 3) Histogram of the exact PCI SNR based on 300 trials. The observed noise data is signal-free.

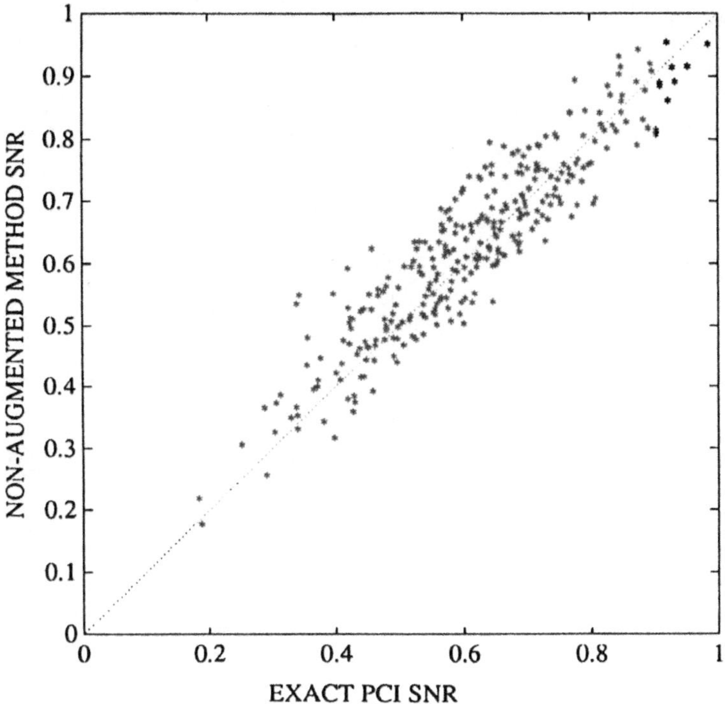

Fig. 4) Scattergram of the non-augmented method SNR vs. exact PCI SNR based on 300 trials. The observed noise data is signal-corrupted.

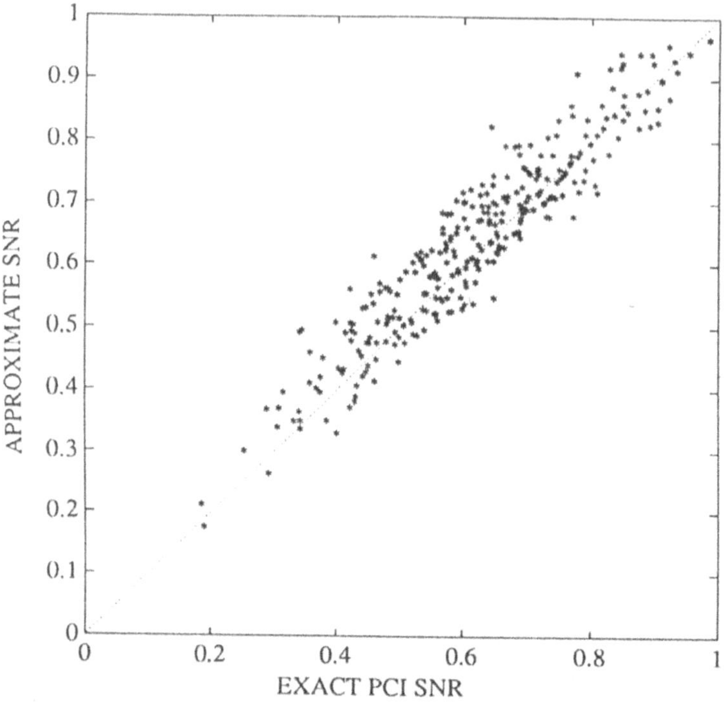

Fig. 5) Scattergram of the approximate SNR vs. exact PCI SNR based on 300 trials. The observed noise data is signal-corrupted.

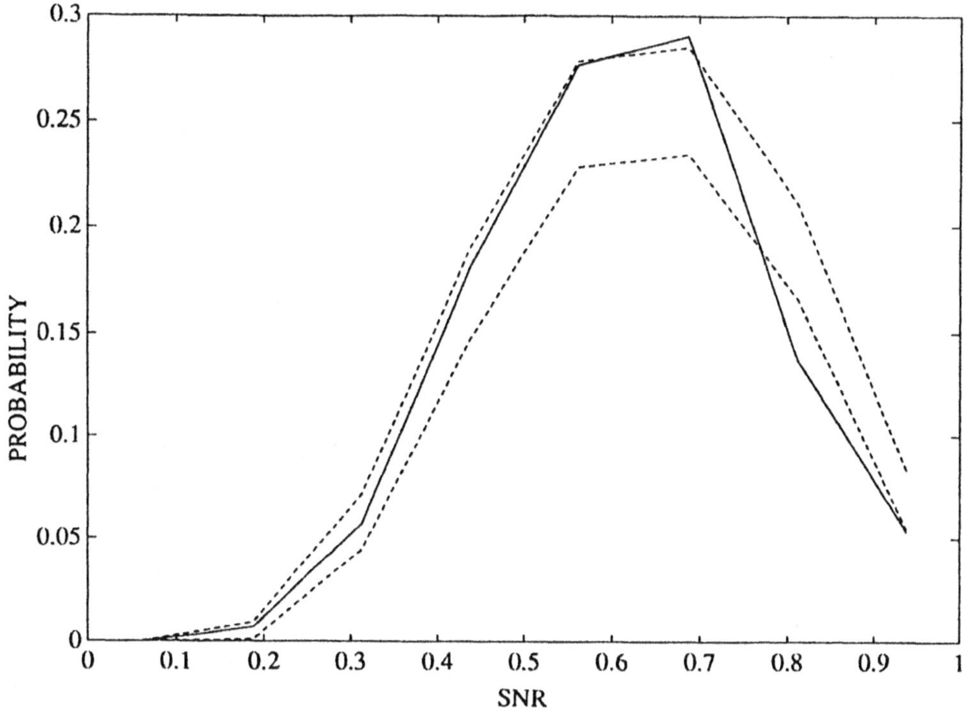

---------- CONFIDENCE BOUND FOR EACH HISTOGRAM BIN BASED ON THE ESTIMATED STANDARD DEVIATION ABOUT THE EXPECTED BIN FREQUENCY USING THE BINOMIAL DISTRIBUTION

Fig. 6) Histogram of the exact PCI SNR based on 300 trials. The observed noise data is signal-corrupted.

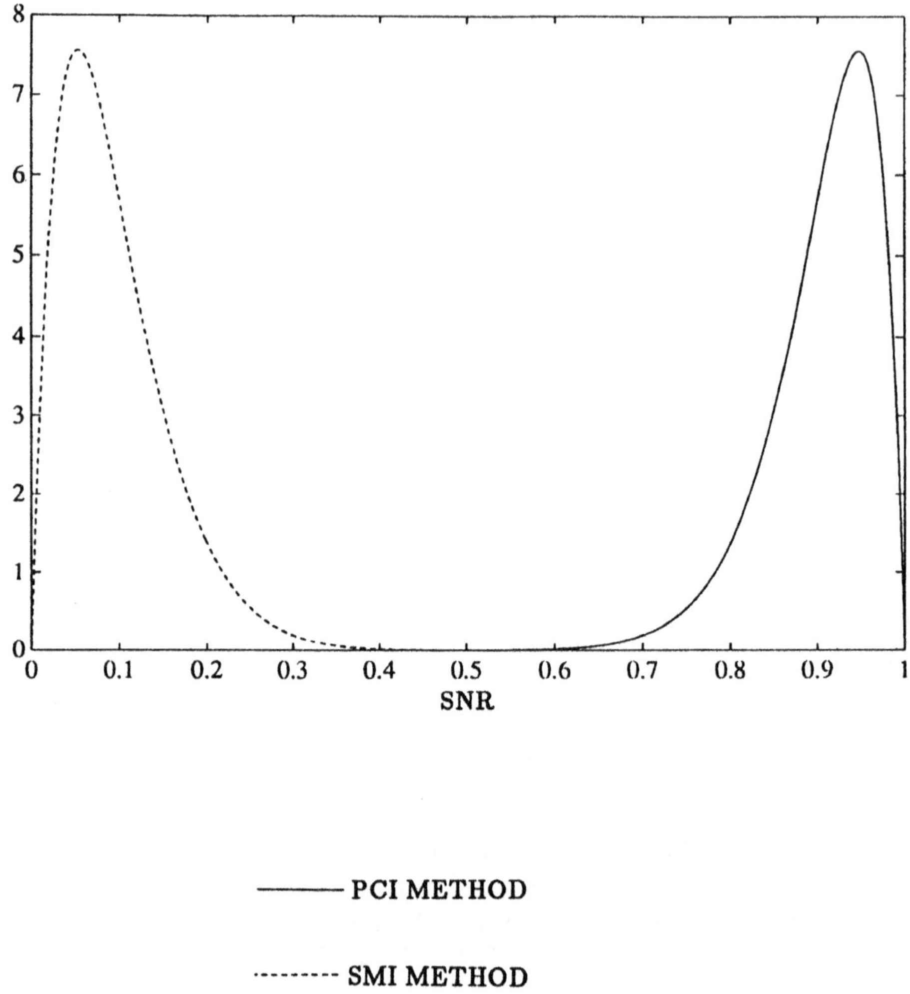

Fig. 7) Theoretical probability density of SNR for the PCI and SMI methods. The observed noise data is signal-free.

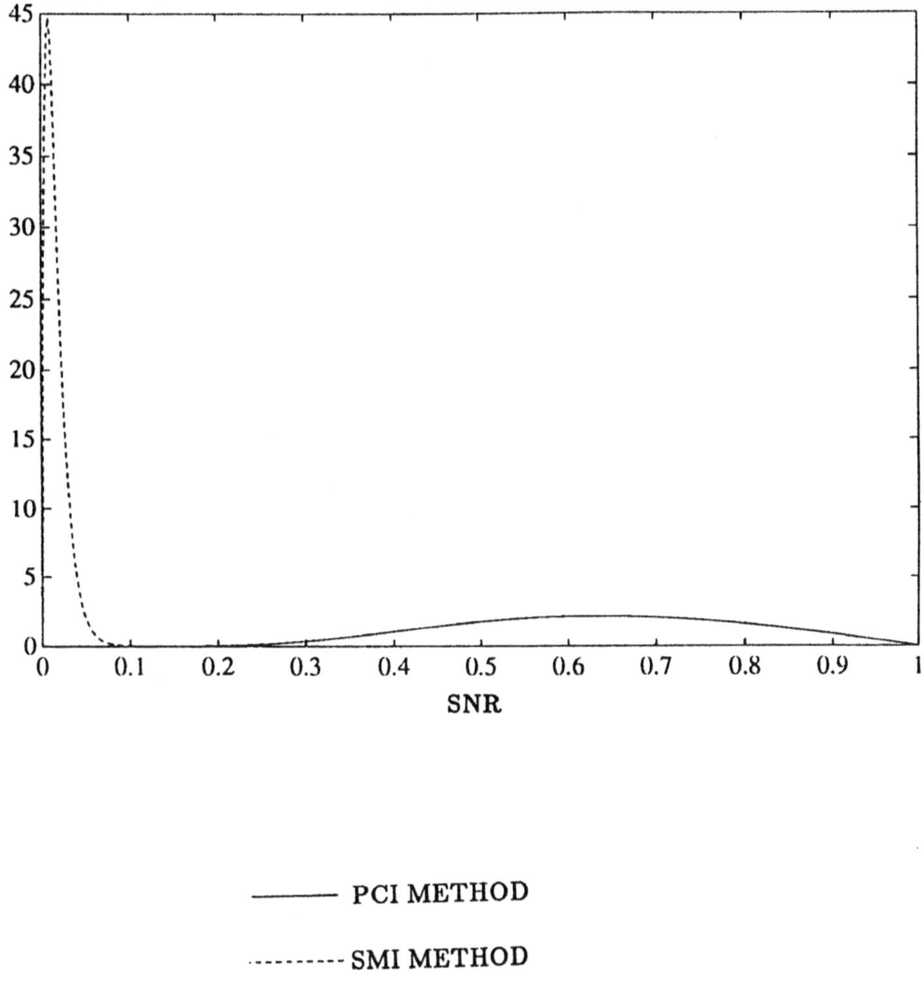

Fig. 8) Theoretical probability density of SNR for the PCI and SMI methods. The observed noise data is signal-corrupted.

Lecture Notes in Control and Information Sciences

Edited by M. Thoma and A. Wyner

Vol. 97: I. Lasiecka/R. Triggiani (Eds.)
Control Problems for Systems
Described by Partial Differential Equations
and Applications
Proceedings of the IFIP-WG 7.2
Working Conference
Gainesville, Florida, February 3-6, 1986
VIII, 400 pages, 1987.

Vol. 98: A. Aloneftis
Stochastic Adaptive Control
Results and Simulation
XII, 120 pages, 1987.

Vol. 99: S. P. Bhattacharyya
Robust Stabilization Against
Structured Perturbations
IX, 172 pages, 1987.

Vol. 100: J. P. Zolésio (Editor)
Boundary Control and Boundary Variations
Proceedings of the IFIP WG 7.2 Conference
Nice, France, June 10-13, 1987
IV, 398 pages, 1988.

Vol. 101: P. E. Crouch,
A. J. van der Schaft
Variational and Hamiltonian
Control Systems
IV, 121 pages, 1987.

Vol. 102: F. Kappel, K. Kunisch,
W. Schappacher (Eds.)
Distributed Parameter Systems
Proceedings of the 3rd International Conference
Vorau, Styria, July 6–12, 1986
VII, 343 pages, 1987.

Vol. 103: P. Varaiya, A. B. Kurzhanski (Eds.)
Discrete Event Systems:
Models and Applications
IIASA Conference
Sopron, Hungary, August 3-7, 1987
IX, 282 pages, 1988.

Vol. 104: J. S. Freudenberg/D. P. Looze
Frequency Domain Properties of Scalar
and Multivariable Feedback Systems
VIII, 281 pages, 1988.

Vol. 105: Ch. I. Byrnes/A. Kurzhanski (Eds.)
Modelling and Adaptive Control
Proceedings of the IIASA Conference
Sopron, Hungary, July 1986
V, 379 pages, 1988.

Vol. 106: R. R. Mohler (Editor)
Nonlinear Time Series and
Signal Processing
V, 143 pages, 1988.

Vol. 107: Y. T. Tsay, L.-S. Shieh, St. Barnett
Structural Analysis and Design
of Multivariable Systems
An Algebraic Approach
VIII, 208 pages, 1988.

Vol. 108: K. J. Reinschke
Multivariable Control
A Graph-theoretic Approach
274 pages, 1988.

Vol. 109: M. Vukobratović/R. Stojić
Modern Aircraft Flight Control
VI, 288 pages, 1988.

Vol. 110: B. J. Daiuto, T. T. Hartley,
S. P. Chicatelli
The Hyperbolic Map and Applications
to the Linear Quadratic Regulator
VI, 114 pages, 1989

Vol. 111: A. Bensoussan, J. L. Lions (Eds.)
Analysis and Optimization
of Systems
XIV, 1175 pages, 1988.

Vol. 112: Vojislav Kecman
State-Space Models of Lumped
and Distributed Systems
IX, 280 pages, 1988

Vol. 113: M. Iri, K. Yajima (Eds.)
System Modelling and Optimization
Proceedings of the 13th IFIP Conference
Tokyo, Japan, Aug. 31 – Sept. 4, 1987
IX, 787 pages, 1988.

Vol. 114: A. Bermúdez (Editor)
Control of Partial Differential Equations
Proceedings of the IFIP WG 7.2
Working Conference
Santiago de Compostela, Spain, July 6–9, 1987
IX, 318 pages, 1989

Vol. 115: H.J. Zwart
Geometric Theory for Infinite
Dimensional Systems
VIII, 156 pages, 1989.

Vol. 116: M.D. Mesarovic, Y. Takahara
Abstract Systems Theory
VIII, 439 pages, 1989

Lecture Notes in Control and Information Sciences

Edited by M. Thoma and A. Wyner

Vol. 117: K.J. Hunt
Stochastic Optimal Control Theory
with Application in Self-Tuning Control
X, 308 pages, 1989.

Vol. 118: L. Dai
Singular Control Systems
IX, 332 pages, 1989

Vol. 119: T. Başar, P. Bernhard
Differential Games and Applications
VII, 201 pages, 1989

Vol. 120: L. Trave, A. Titli, A. M. Tarras
Large Scale Systems:
Decentralization, Structure Constraints
and Fixed Modes
XIV, 384 pages, 1989

Vol. 121: A. Blaquière (Editor)
Modeling and Control of Systems
in Engineering, Quantum Mechanics,
Economics and Biosciences
Proceedings of the Bellman Continuum
Workshop 1988, June 13–14, Sophia Antipolis, France
XXVI, 519 pages, 1989

Vol. 122: J. Descusse, M. Fliess, A. Isidori,
D. Leborgne (Eds.)
New Trends in Nonlinear Control Theory
Proceedings of an International
Conference on Nonlinear Systems,
Nantes, France, June 13–17, 1988
VIII, 528 pages, 1989

Vol. 123: C. W. de Silva, A. G. J. MacFarlane
Knowledge-Based Control with
Application to Robots
X, 196 pages, 1989

Vol. 124: A. A. Bahnasawi, M. S. Mahmoud
Control of Partially-Known
Dynamical Systems
XI, 228 pages, 1989

Vol. 125: J. Simon (Ed.)
Control of Boundaries and Stabilization
Proceedings of the IFIP WG 7.2 Conference
Clermont Ferrand, France, June 20–23, 1988
IX, 266 pages, 1989

Vol. 126: N. Christopeit, K. Helmes
M. Kohlmann (Eds.)
Stochastic Differential Systems
Proceedings of the 4th Bad Honnef Conference
June 20–24, 1988
IX, 342 pages, 1989

Vol.127: C. Heij
Deterministic Identification
of Dynamical Systems
VI, 292 pages, 1989

Vol. 128: G. Einarsson, T. Ericson,
I. Ingemarsson, R. Johannesson,
K. Zigangirov, C.-E. Sundberg
Topics in Coding Theory
VII, 176 pages, 1989

Vol. 129: W. A. Porter, S. C. Kak (Eds.)
Advances in Communications and
Signal Processing
VI, 376 pages, 1989.

Vol. 130: W. A. Porter, S. C. Kak,
J. L. Aravena (Eds.)
Advances in Computing and Control
VI, 367 pages, 1989

Vol. 131: S. M. Joshi
Control of Large Flexible Space Structures
IX, 196 pages, 1989.

Vol. 132: W.-Y. Ng
Interactive Multi-Objective Programming
as a Framework for Computer-Aided Control
System Design
XV, 182 pages, 1989.

Vol. 133: R. P. Leland
Stochastic Models for Laser Propagation
in Atmospheric Turbulence
VII, 145 pages, 1989.

Vol. 134: X. J. Zhang
Auxiliary Signal Design in Fault
Detection and Diagnosis
XII, 213 pages, 1989

Vol. 135: H. Nijmeijer, J. M. Schumacher (Eds.)
Three Decades of Mathematical System Theory
A Collection of Surveys at the Occasion of the
50th Birthday of Jan C. Willems
VI, 562 pages, 1989

Vol. 136: J. Zabczyk (Ed.)
Stochastic Systems and Optimization
Proceedings of the 6th IFIP WG 7.1
Working Conference,
Warsaw, Poland, September 12–16, 1988
VI, 374 pages. 1989

Lecture Notes in Control and Information Sciences

Edited by M. Thoma and A. Wyner

GPSR Compliance

The European Union's (EU) General Product Safety Regulation (GPSR) is a set of rules that requires consumer products to be safe and our obligations to ensure this.

If you have any concerns about our products, you can contact us on ProductSafety@springernature.com

In case Publisher is established outside the EU, the EU authorized representative is:

Springer Nature Customer Service Center GmbH
Europaplatz 3
69115 Heidelberg, Germany

Batch number: 10704763

Printed by Printforce, the Netherlands